Advances in Dynamical Systems Theory, Models, Algorithms and Applications

Edited by Bruno Carpentieri

Published in London, United Kingdom

IntechOpen

Supporting open minds since 2005

Advances in Dynamical Systems Theory, Models, Algorithms and Applications
http://dx.doi.org/10.5772/intechopen.92486
Edited by Bruno Carpentieri

Contributors
Alberto Postiglione, Giancarlo Nota, Lionel Merveil Merveil Anague Tabejieu, Blaise Roméo Nana NBendjo, Giovanni Filatrella, Lal Mohan Saha, Mrinal Kanti Das, Elena Karachanskaya, Michel Moreau, Bernard Gaveau, Adela Ionescu

First published in London, United Kingdom, 2021 by IntechOpen
IntechOpen is the global imprint of INTECHOPEN LIMITED, registered in England and Wales, registration number: 11086078, 5 Princes Gate Court, London, SW7 2QJ, United Kingdom
Printed in Croatia

British Library Cataloguing-in-Publication Data
A catalogue record for this book is available from the British Library

Additional hard and PDF copies can be obtained from orders@intechopen.com

Advances in Dynamical Systems Theory, Models, Algorithms and Applications
Edited by Bruno Carpentieri
p. cm.
Print ISBN 978-1-83969-123-2
Online ISBN 978-1-83969-124-9
eBook (PDF) ISBN 978-1-83969-125-6

Meet the editor

Bruno Carpentieri obtained a Laurea degree in Applied Mathematics from Bari University in 1997 and a Ph.D. in Computer Science from the Institut National Polytechnique de Toulouse (INPT), France. After some post-doctoral experiences, Dr. Carpentieri served as an assistant professor at the Bernoulli Institute for Mathematics, Computer Science and Artificial Intelligence, University of Groningen, the Netherlands, and as a reader in Applied Mathematics at Nottingham Trent University, England. Since May 2017 he has been an associate professor of Applied Mathematics, Faculty of Computer Science, Free University of Bozen-Bolzano, Italy. His research interests include applied mathematics, numerical linear algebra, and high-performance computing. Dr. Carpentieri served as a member of several scientific advisory boards in computational mathematics. He is an editorial board member of the Journal of Applied Mathematics, an editorial committee member of Mathematical Reviews (American Mathematical Society) and a reviewer of about thirty scientific journals in numerical analysis. He has co-authored about fifty publications in peer-reviewed scientific journals.

Contents

Preface

This book examines some problems and current trends relating to the theory of dynamical systems. It also discusses some significant advances made in this field in the last several years, including new models, computer algorithms, and applications in various scientific areas.

The book includes six chapters. Chapter 1 presents an innovative methodology that integrates physical devices and machinery with text mining technologies to identify anomalous behaviors, even of minimal entity, rarely perceived by other strategies in a machine tool. Chapter 2 presents a vibrations analysis of elastic beams based on the constitutive equation of Kelvin-Voigt type, which contains fractional derivatives of real order. Chapter 3 discusses the emergence of chaos and complex behavior in real and physical systems, reviewing problems such as the complexity of Child's swing dynamics, chaotic neuronal dynamics, complex food-web dynamics, financial models, and others. Chapter 4 considers the invariant method theory for stochastic systems with strong perturbations, representing modern approaches to describe dynamical systems having a set of invariant functions. Chapter 5 presents a study of a mesoscopic stochastic process derived from deterministic dynamics studied by Kolmogorov and others in the stationary case, and extends their methods and some of their results considering the non-stationary process, which stems from a non-invariant initial measure. Finally, Chapter 6 introduces recent developments on the stability analysis of dynamical systems using the powerful tool of control Lyapunov functions.

Researchers, engineers, and graduate students in both pure and applied mathematics may benefit from the chapters collected in this volume. We express appreciation to IntechOpen for their professional support and to Author Service Manager Mr. Josip Knapić for his tireless help in the preparation of this book.

Bruno Carpentieri
Free University of Bozen-Bolzano (Faculty of Computer Science),
Bolzano, Italy

Chapter 1

Text Mining for Industrial Machine Predictive Maintenance with Multiple Data Sources

Giancarlo Nota and Alberto Postiglione

Abstract

This paper presents an innovative methodology, from which an efficient system prototype is derived, for the algorithmic prediction of malfunctions of a generic industrial machine tool. It integrates physical devices and machinery with Text Mining technologies and allows the identification of anomalous behaviors, even of minimal entity, rarely perceived by other strategies in a machine tool. The system works without waiting for the end of the shift or the planned stop of the machine. Operationally, the system analyzes the log messages emitted by multiple data sources associated with a machine tool (such as different types of sensors and log files produced by part programs running on CNC or PLC) and deduces whether they can be inferred from them future machine malfunctions. In a preliminary offline phase, the system associates an alert level with each message and stores it in a data structure. At runtime, three algorithms guide the system: pre-processing, matching and analysis: Preprocessing, performed only once, builds the data structure; Matching, in which the system issues the alert level associated with the message; Analysis, which identifies possible future criticalities. It can also analyze an entire historical series of stored messages The algorithms have a linear execution time and are independent of the size of the data structure, which does not need to be sorted and therefore can be updated without any computational effort.

Keywords: industrial machine tool, predictive maintenance, log message, text mining, efficient algorithm

1. Introduction

The concept of Predictive Maintenance [1–4] foresees the carrying out of maintenance activities before the equipment failure. The primary goal of predictive maintenance is to reduce the frequency of equipment failures by preventing the failure before it actually occurs [5]. This strategy helps to minimize breakdown costs and downtime (loss of production) and increase product quality, well known thanks to [6] and recently reiterated by [4]. Obviously [7] predictive maintenance is different from corrective maintenance, as action will be taken here to "anticipate" the error before it actually occurs. Predictive maintenance is primarily about detecting hidden and potential failures. It does not replace, but joins the Preventive Maintenance in the strict sense, which is linked to the execution of a specific protocol (often agreed with the machine manufacturer) intended to periodically check, or after a certain amount of work, the state of use of the machine, without any signs of behavioral anomalies having actually occurred. According to [8], if the maintenance strategy only involves interventions that react to failures, the maintenance costs are relatively low but the losses could be high. If preventive and predictive maintenance is introduced, maintenance costs increase: for example, some activities must be carried out using overtime, detectors for predictive maintenance are introduced, time is dedicated to training activities for operators and maintenance workers. Some clues for defining the algorithms on which the cyber-physical system presented here is based derive from research in the field of computational linguistics, which have appeared on papers presented at various important international conferences [9–14].

Here we present a discrete dynamic system, based on events represented as textual messages to which an alert level has been associated and whose data structure is a graph. The system is dynamic because the data structure adjusts itself without additional computational costs if a new message is issued by the machine, which has never appeared before and that is not yet included in the set of known messages. The new message must in any case be validated by a human expert, who must associate the message with an adequate alert level.

This paper is structured as follows: in section 2 we present the main characteristics of the data emitted by the data sources associated with an industrial machine tool, and how they are represented in the system. In section 3 we present the system. In particular, in section 3.1 we present the model for a machine tool, in section 3.2 we give an overview of the system, while in section 3.3 and in section 3.4 we present the two main phases of the system: pre-processing, to be performed only once, and runtime; in section 3.5 and in section 3.6 we present the main algorithms and their theoretical performances; in section 4 we report a brief summary of the results of a prototype of the system applied to a simple, but real, case study. The paper ends in section 5 with some conclusions.

2. Industrial machine tool data

In the following, the term "machine data" defines an information message emitted by any "data source" associated with a machine tool (for example a sensor), which concerns an event that occurred during the activity of the machine itself. Machine data are the log files emitted by the machine itself, but also all data emitted by external entities such as event sensors, sensors for speed, temperature, acceleration and so on.

In order to reconstruct in detail the "history" of the messages, machine tool data must be raw and immutable (as opposed to classical structured and aggregated Relational Databases data). Data is never deleted or updated (except in very rare cases, for example to comply with regulations), it is only added. The main disadvantage of this management is that the stored data tends to become very large. From this point of view we can talk about big data. To avoid misinterpretation of the data, they are not stored in a free format, but are "semantically normalized", that is, remodeled in a standard format, even if not strictly structured.

Machine tools data are represented through the "fact-based model" (see **Figure 1**): a graph where each message corresponds to a single fact. In this graph, each node corresponds to a machine tool data source entity (e.g. Sensor Y on machine M) and each arc can represent information about an entity (dashed connecting line) or a relationship between two entities (continuous line). Each single message is identified by its timestamp plus the identification of the machine (M) and the identification of the entity that emits the message (Sensor Y).

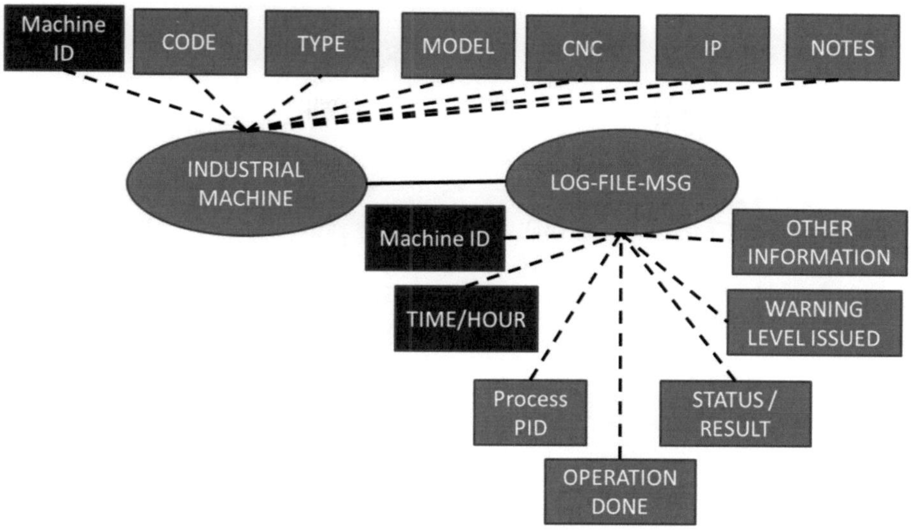

Figure 1.
A simple fact-based model for a standard log file.

Physically, we assume that each source message emitted by a machine tool is in a semi-structured text format (a sort of simplified JSON): that is, it is a succession of text fields, divided by a field terminator. This provides simplicity and flexibility with the greatest possible space savings. It is possible to store anything within the main dataset, as long as each data has the same information placed in the same order, with the same data type format; otherwise it will be necessary to carry out a pre-processing of the source messages before storing them, downloading the data that do not conform to the expected format in a separate archive.

3. System description

3.1 The model for machine tool

Manufacturing systems organize machine tools, material handling equipment, inspection equipment, and other manufacturing assets in a variety of layouts. With the advent of Industry 4.0 technologies, these manufacturing resources can be networked, and cyber-physical manufacturing systems can be implemented that integrate hardware and software with mechanical and electrical systems designed for a specific purpose. The model of **Figure 2** shows a representation of the main components of a generic industrial machine from the point of view of data collection and is sufficiently general to be applied to many industrial machines that can be part of process, cellular and line layouts.

The generic machine Mi receives from the network a command message known as part program. It is a set of detailed step-by-step instructions executed by the CONTROL system (CNC, PLC) that direct the actions implemented by the actuators. The actuators act as transducers changing a physical quantity, typically electric current, into another type of physical quantity such as rotational speed of an electric motor. During the execution of a part program, the machine tool operates and the CONTROL system produces a log containing data about the executed instructions as well as control messages that indicate some particular machine state. In the meantime, execution data captured by the sensors are sent via the communication

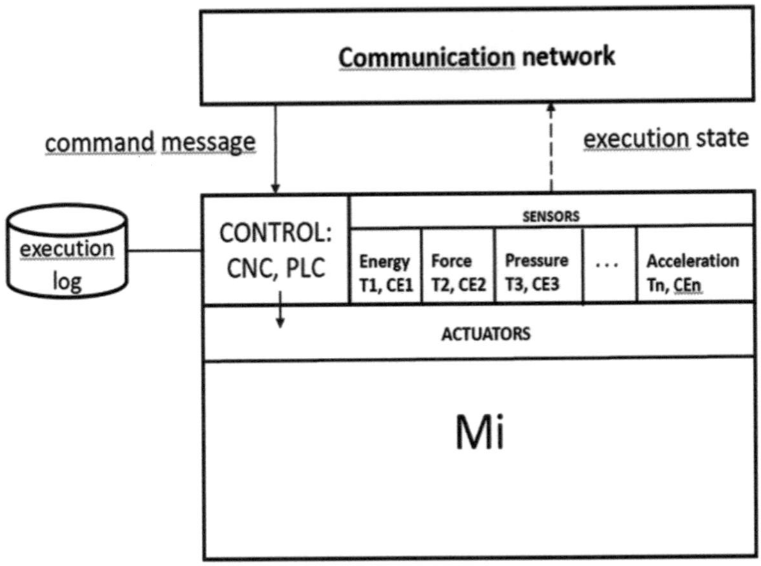

Figure 2.
A simple model for machine tool data collection.

network to the factory control system. Here, the feedback loop is closed, and feedback actions can eventually be taken.

The generic sensor is represented by its two constituent parts: the transducer (Ti) and the control electronics (CEi); this distinction is useful because even if the sensor is a unified whole, the transducer and the control electronics can be placed into different area of the machine tool. For example, the accelerometer transducer can be placed on the spindle while its control electronics is usually placed within the cabinet together with other control subsystems.

3.2 System functionality

The event-based dynamic cyber-physical system proposed here integrates physical devices with advanced Text Mining analytical technologies and is very general, therefore adaptable to any production domain. It needs the ontology of the messages emitted by the data sources of a machine tool during its normal operation and is able to intercept its (even slightly) anomalous behaviors, which allow to evaluate whether it is moving towards a failure state such as to require the activation of safeguard procedures.

The system consists of a design pre-processing phase to create the main message ontology and which is performed only once for each data source (for example a sensor), and an algorithmic runtime phase. Each message correspond to an event of the data source, has the form of a text messages and has associated an adequate alert level.

3.3 Pre-processing design phase

The pre-processing phase is performed only once for each data source and allows you to create the initial ontology of the messages. It consists of four main steps.

In the first step of the pre-processing phase, for each data source the set of all messages that it can emits is identified and normalized. In particular, for each industrial machine and for each data source associated with that machine, the possible types of messages that can be issued must be identified; then, these data must be interpreted, choosing only the pertinent ones; finally, the data must be semantically normalized, to adapt it to a common data semantics.

In the next step of the pre-processing phase, the system associates an alert level, based on a chromatic scale, to each of the messages coming from the previous step. The levels are as follows:

White Level (no problem).

Yellow Level (warning): There have been sporadic, but not serious, anomalies and the system is able to continue without problems.

Orange Level (serious): some serious anomalies, or some cluster of anomalies, have been found in the system, which could affect the future of work.

Red Level (very serious): Very serious anomalies were found in the system, capable of affecting, in the very near future, the production activity.

Black Level (immediate stop): the system runs the immediate risk of irreparable equipment or product failures.

In the third step of the pre-processing phase, to be carried out only if there are several data sources associated with the machine tool, composite messages are created, obtained by composing the messages with at least an yellow alert level, a composite message. The order in which messages are juxtaposed is predefined and must be the same as it will appear at runtime. At this point, a general alert level is associated with each composite message.

Finally, a digital data structure is built containing all composite messages: a text dictionary, in which each line contains a composite message and the relative general alert level.

Steps 1 and 2 must be performed for each of the data source entities associated with a specific machine tool. Identifying alert levels requires the help of the machine tool expert, thus steps 2 and 3 need the help of a machine tool expert.

The whole data design process is reported in **Figure 3**.

3.4 Runtime phase

In the runtime phase, the real data coming from a machine tool are collected and normalized according to the specifications identified in the pre-processing phase. Then, all messages with a compatible timestamp are aggregated into a line of text to

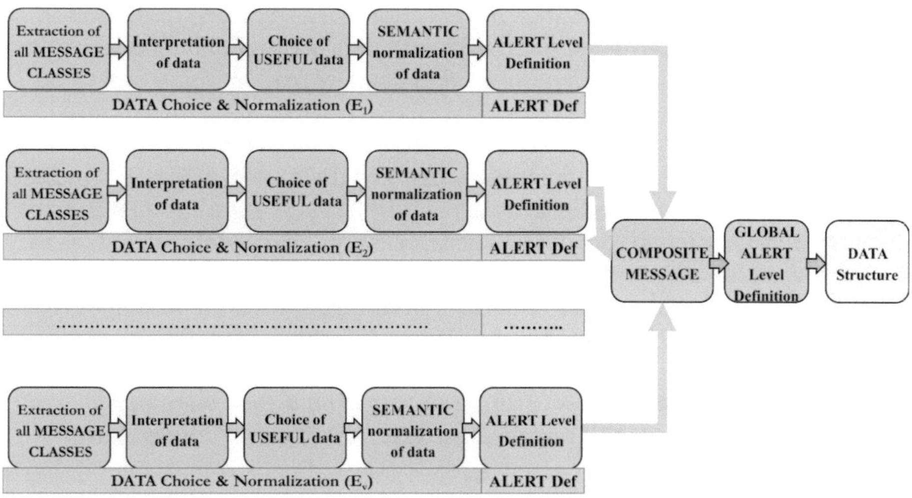

Figure 3.
Data design process.

form a single composite message that has the same layout as the composite messages in the dictionary. The layout of a composite message from a machine tool and n Data Sources, is as follows:

Machine ID + Timestamp.
+ DS_1 ID + DS_1 Message.
+ DS_2 ID + DS_2 Message.
+ ...
+ DS_n ID + DS_n Message.

The algorithms look up this compound message in the dictionary of all compound messages and extract its global alert level. They identify all occurrences of all messages in the text, even if they are overlapped (partially or completely).

At this point the system analyzes the alert levels found, their frequency and, possibly, their relative positioning both in time and in space, and predicts possible future malfunctions.

The system implements different types of analysis algorithms. The simplest algorithm catalogs the entire cluster of messages and identifies the warning levels that appear most frequently. A more sophisticated algorithm analyzes smaller homogeneous clusters of these composite messages. Another algorithm identifies the presence of sub-clusters, even small ones, which indicate a possible future malfunction of the machine.

The software is written in DELPHI (Visual Development Environment), is highly parameterized (to allow quick modification) and is composed of about 2,000 lines of code structured in 30 operating modules, different methods for managing the interface and uses a large number of predefined libraries.

3.5 Detailed algorithms description

The system at runtime is fast and accurate in identifying possible anomalous situations of a machine tool. It consists of a pre-processing step (to be performed only once, when the program starts) which has the purpose of building the entire data structure (the dictionary of all messages with their relative alert level) and an actual processing step.

The system is independent on the dictionary size. It reads the input text, consisting of one or more messages, one character at a time and scrolls through a finite automaton; when it encounters a final state (corresponding to the end of a message), it emits the alert level associated with that message. The number of state transitions is proportional to the number of characters read in input.

The steps are:

Pre-processing. The system builds the data structure that the algorithms will use when running. Pre-processing only needs to be done once, at startup. In particular, the system constructs, in the central memory, a finite state automaton from all the elements of the dictionary, and its behavior is completely determined by a (small) set of states and by some simple functions.

Matching. The finite state automaton continuously reads the messages from the input log file, character by character. When it reaches a final state, the automaton shows the alert level of the message corresponding to this state. The algorithm is able to identify all occurrences of all messages, even if they were partially or totally overlapping. Lines with no messages (i.e. with "blank" content) are ignored.

Analysis. In this step, the system analyzes a set of alert levels identified in the previous phase. Normally, it analyzes a group of alarms, usually those issued in a period; then, based on the frequency and "severity" of the alerts extracted, it makes a classification of the health of the machine. The system warns if the analysis of the message cluster indicates a possible future malfunction of the machine tool. The system can also provide an immediate response if a single alert is very critical.

3.6 Performance analysis of algorithms

The pre-processing step (i.e. the construction of the FA) requires a time linearly proportional to the sum of the lengths of all the messages present in the dictionary, i.e. the total number of characters in the dictionary.

The matching algorithm for a set of textual log messages with a total length of k characters requires n state transitions ($n \leq 2k$). Therefore, the analysis of a message takes a linear time with respect to the number of characters, and this is a lower limit, whatever the algorithm used among those who read the message character by character, since all characters of the message must be read.

Algorithms with an approach different from the one proposed here, should refer to the method that considers a whole input message as a single entity. These methods look for a message in the set of all possible messages (the dictionary). From the literature [15], we know that no algorithm in that class can use less than $O(logn)$ steps to look up a single message in a dictionary of n messages.

Therefore, the algorithms used here are independent of the size of the dictionary, which can be as large as we want without worsening the search time, while the classical algorithms are dependent on the size of the dictionary: the larger it is, the more time it takes to search for a message within it.

The differences are even greater by admitting the possibility of editing the dictionary. With our approach, the dictionary does not need to be sorted, so adding, changing, or deleting one or more dictionary entries is done at virtually no computation cost. On the other hand, with the classical approach, the dictionary must be kept orderly. Therefore, in case of cancelation, insertion or modification even of a single entry, the dictionary must be reordered, and this costs at least $O(logn)$ operations, possibly with physical movement of the entries from one memory area to another.

Another advantage of our approach is of a technical nature: the algorithms are all executed in central memory, while a classical method largely uses secondary memories (which are much slower by several orders of magnitude).

In addition, each log message (dictionary entry) is made up of (many) words and other non-alphabetic symbols and this means that the dictionary size can be very large and, furthermore, the search algorithms need the messages to be well delimited and not superimposed, while our approach is able to identify even totally overlapping or non-delimited messages and present within plain text sentences.

4. A simple case study

We have also implemented a prototype of the cyber-physical system presented in this paper. In this section we report the results of some preliminary tests we conducted on some data coming from a machine tool currently operating in an important company, operating in Southern Italy, which produces metal molds for other national and international companies.

In particular, we analyzed a log file relating to a period of 194 hours of continuous work of a machine tool, from 4:25 on 13 / Feb / 2019 to 21:34 on 20 / Feb / 2019, in which it issued approximately 300,000 log messages.

In the preliminary testing phase, we used data from a single data source from a single machine tool, then we considered simple, not composite messages, because our initial interest was to verify the feasibility of algorithmic message search in ontology.

4.1 Case study data description

A log is the sequential and chronological recording of the operations and events coming from a specific industrial machine tool. Log messages are stored in text format in one or more files, one record per line with each line containing only one message. Generally, these registrations are done in an automated way. Each record stores everything that happens on the machine, so a log file holds both information about normal machine operation and about errors and problems or even slight deviations from the norm.

This section shows the actual data relating to the case study examined, but the information contained in a log file of a generic machine tool is approximately the same, regardless of the machine; at most it could slightly change the data format in each individual field or their mutual position in the log file.

Therefore, the case study presented here is representative of many industrial machines.

Here, there is an example of the industrial machine log file from the case study.

13/02/2019 04:25:24;MSG_SYS;Scrive;Fine corsa asse.., Y+;

13/02/2019 04:25:26;MSG_SYS;Scrive;E 2034:indice variabile errato in riga PLC.. 4239;

13/02/2019 04:25:28;MSG_SYS;Scrive;Fine corsa asse.., Y+;

13/02/2019 04:25:30;MSG_SYS;Scrive;E 2034:indice variabile errato in riga PLC.. 4239;

The log file has four blocks of useful data and a fifth block that is not useful for analysis. Each field is separated from the next by the semicolon symbol (;). The essential information contained in the first four blocks of the first row (the one in bold) of the example above are:

13/02/2019 04:25:24;	**Date/time of recording of the event**
MSG_SYS;	Process PID, i.e. the identification of the running process,
Scrive;	The operation done
Fine corsa asse.., Y+;	The status or result of the execution of the event

As described in the following section, we have extracted all the different messages from the 300,000 input lines; therefore we have assigned, with the assistance of the machine tool technician, an alert level to each single different message.

A single log record stored in the dictionary is a line containing the following information, separated by an hashtag:

"Machine ID" + Date/Time # Process PID # Oper. Done # Result # "Warning Level" (1).

Machine ID is a code that allows to identify the single machine. Since the same message can be emitted several times on the same machine, but never at the same instant, each message is identified by ID + Date/time. Each message corresponds to a single fact and, associated with each message, there is the level of warning issued. By adding the ID of the data source to the message, it is possible to integrate messages from different data sources that insist on the same machine.

4.2 System prototype run and results

In the preliminary phase, we had to identify the different types of messages emitted by the machine for which we had the log files. We analyzed 300 log files, each containing 1000 messages, from 4:25 on feb/13/2019 to 21:34 on feb/20/2019.

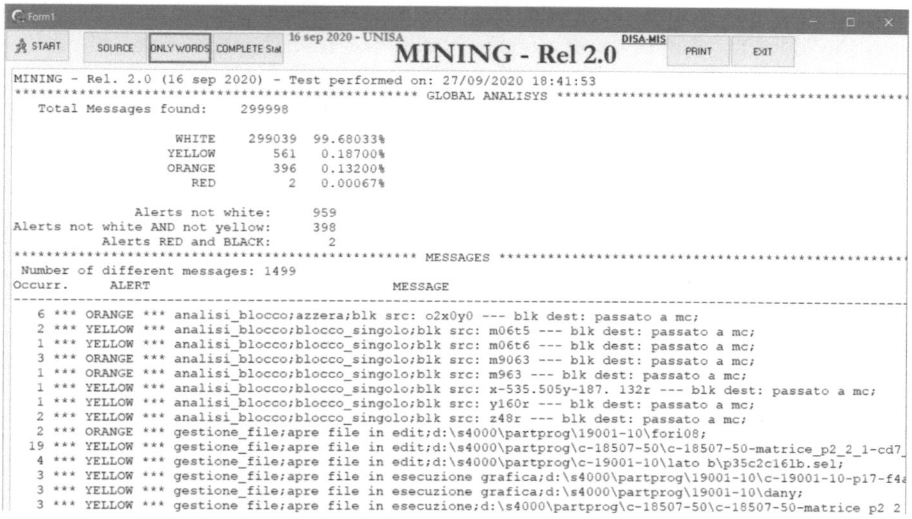

Figure 4.
One week summary of log message analysis.

Of the 300,000 messages (the non-empty ones are actually 299,998), which we have inserted in a single file, those with different "semantic" content are 1,499. To achieve this we have eliminated every message that does not have useful information and the first 20 characters (date and time) of each remaining line and then we have identified all the different messages. At the end of this first pre-processing phase, we have a file that contains all and only 1,499 different messages.

In the next step, with the assistance of the machine tool technician, we assigned an ALERT level to each of the 1,499 different messages. The alert levels are ordered according to the increasing level of severity: white, yellow, orange, red and black.

The final document contains the associations between the 1,499 messages and the related warning level. We report here the screen of the test sent by us on the log file and the first part of the list of messages encountered with the relative multiplicity and alerts: In **Figure 4** there is the final report of a week's analysis of log messages from a real industrial machine.

During this week there were three times when groups of messages occurred that required the supervision of experienced personnel, but none of these alarms turned into a request to stop the machine. Two interventions were due to a red alert and another due to an anomalous aggregation of orange and yellow alerts in a short period. At the end of the week, during a planned machine downtime, some adjustments were made, suggested by the presence of some clusters of messages relating to a slight deviation of the machine performance, compared to the estimated one.

Since the messages can be analyzed in real time, if clusters of "dangerous" messages occurr during the operation of the machine, the machine experts would been able to intervene in time to prevent irreparable damage. Note that the message cluster can also be formed by several messages of mild severity issued, however, in an anomalous configuration or within very short times.

5. Conclusions

We have presented here an innovative methodology and an associated fast and efficient software system prototype, for the algorithmic prediction of industrial machine tools malfunctions, adaptable to any type of company. It integrates machinery and physical devices with the analytical technologies of Text Mining and allows the identification of anomalous behavior of a machine tool, even of minimal entity, rarely perceived by other strategies.

The system performs its analysis without waiting for the end of the shift or a machine stop. After recognition, it can initiate automatic safeguard procedures, call a human expert, or schedule some minor tuning operations. The system works without waiting for the shift to end or the machine to stop.

The algorithms require linear execution time on the number of input characters, run on a data structure completely on RAM and are independent on the data structure size, which can be modified without actual computational costs, as it is not sorted. A classic approach, on the other hand, requires searching for a message in a set of possible messages using efficient algorithms, which work largely on secondary memories and depend on the size of the data structure that need to be, necessary, sorted, and therefore it takes time, not irrelevant, to add, modify or delete an entry in it, possibly by physically moving items from one memory area to another.

Last, but not least, is the fact that a classic approach is inadequate because a log message is made up of many words and others non-alphabetic symbols and the data structure size could be very large.

We believe that this approach can bring significant competitive advantages to a company in which the effective and precise predictive analysis of machine tools is a necessity to be pursued by spending as little time as possible, obtaining as precise a result as possible, limiting false recognition errors, as much as possible.

Author details

Giancarlo Nota† and Alberto Postiglione*†
Department of Business Sciences - Management and Innovation Systems (DISA-MIS), University of Salerno, Salerno, Italy

*Address all correspondence to: apostiglione@unisa.it

† These authors contributed equally.

References

[1] Löfsten H. Management of industrial maintenance – economic evaluation of maintenance policies. International Journal of Operations & Production Management. 1999 Jul;19(7):716–737. Available from: https://doi.org/10.1108%2F01443579910271683.

[2] Ahmad R, Kamaruddin S. An overview of time-based and condition-based maintenance in industrial application. Computers & Industrial Engineering. 2012 Aug;63(1):135–149. Available from: https://doi.org/10.1016%2Fj.cie.2012.02.002.

[3] Carneiro D, Nunes D, Sousa C. A Decision-Support System for Preventive Maintenance in Street Lighting Networks. In: 18th International Conference on Hybrid Intelligent Systems (HIS 2018), Porto, Portugal, December 13–15. vol. 923. Springer International Publishing, AISC series; 2019. p. 272–281.

[4] Vilarinho S, Lopes I, Oliveira JA. Preventive Maintenance Decisions through Maintenance Optimization Models: A Case Study. Procedia Manufacturing. 2017 Jun;11:1170–1177.

[5] Szczepaniak M, Trojanowska J. Preventive Maintenance System in a Company from the Printing Industry. In: et al IV, editor. Advances in Design, Simulation and Manufacturing II. DSMIE 2019, Lecture Notes in Mechanical Engineering. Springer International Publishing; 2019. p. 351–358.

[6] Usher JS, Kamal AH, Syed WH. Cost optimal preventive maintenance and replacement scheduling. IIE Transactions. 1998 Dec;30(12):1121–1128.

[7] Li J, Tao F, Cheng Y, Zhao L. Big data in product lifecycle management. The International Journal of Advanced Manufacturing Technology. 2015;81: 667–684.

[8] DeFelice F, Petrillo A, Monfreda S. Improving Operations Performance with World Class Manufacturing Technique: A Case in Automotive Industry. In: Operations Management. InTech; 2013. p. 1.

[9] Postiglione A, Monteleone M. Towards automatic filing of corpora. In: Proceedings of 18'eme COLLOQUE INTERNATIONAL "Lexique et Grammaires Comparçs", Parco Scientifico e Tecnologico di Salerno e delle aree interne della Campania, Salerno, Oct 6–9; 1999. p. 1–9.

[10] Elia A, Vietri S, Postiglione A, Monteleone M, Marano F. Data mining modular software system. In: Arabnia HR, Marsh A, Solo AMG, editors. SWWS2010 - Proceedings of the 2010 International Conference on Semantic Web & Web Services. Las Vegas, Nevada, USA, Jul. 12–15. CSREA Press; 2010. p. 127–133.

[11] Elia A, Postiglione A, Monteleone M, Monti J, Guglielmo D. CATALOGA: a software for semantic and terminological information retrieval. In: Akerkar R, editor. Proceedings of the International Conference on Web Intelligence, Mining and Semantics, WIMS 2011, Sogndal, Norway, May 25–27, 2011. ACM Press; 2011. p. 11.

[12] Elia A, Postiglione A, Monteleone M. CATALOGA: a software for semantic-based terminological data mining. In: 1st International Conference on Data Compression, Communication and Processing, IEEE, Palinuro (SA), June 21–24. IEEE Computer Society; 2011. p. 153–156. Available from: http://www.computer.org/csdl/proceedings/ccp/2011/4528/00/index.html.

[13] Postiglione A, Monteleone M. Semantic-based bilingual text-mining. In: Second International Conference on

Data Compression, Communication, Processing and Security (CCPS 2016), September 22–23, Cetara (SA); 2016. p. 1–4.

[14] Postiglione A, Monteleone M. A Linguistic Semantic Text-Mining for Multiword Units. In: Roni R, editor. MANTUA HUMANISTIC STUDIES. vol. XII. Mantua Humanistic Studies, Mantova:Mantova: Universitas Studiorum; 2020. p. 445–459.

[15] Cormen TH, Leiserson CE, Rivest RL, Stein C. Introduction to Algorithms, Third Edition. 3rd ed. The MIT Press; 2009.

Chapter 2

Vibrations of an Elastic Beam Subjected by Two Kinds of Moving Loads and Positioned on a Foundation having Fractional Order Viscoelastic Physical Properties

Lionel Merveil Anague Tabejieu,
Blaise Roméo Nana Nbendjo and Giovanni Filatrella

Abstract

The present chapter investigates both the effects of moving loads and of stochastic wind on the steady-state vibration of a first mode Rayleigh elastic beam. The beam is assumed to lay on foundations (bearings) that are characterized by fractional-order viscoelastic material. The viscoelastic property of the foundation is modeled using the constitutive equation of Kelvin-Voigt type, which contain fractional derivatives of real order. Based to the stochastic averaging method, an analytical explanation on the effects of the viscoelastic physical properties and number of the bearings, additive and parametric wind turbulence on the beam oscillations is provided. In particular, it is found that as the number of bearings increase, the resonant amplitude of the beam decreases and shifts towards larger frequency values. The results also indicate that as the order of the fractional derivative increases, the amplitude response decreases. We are also demonstrated that a moderate increase of the additive and parametric wind turbulence contributes to decrease the chance for the beam to reach the resonance. The remarkable agreement between the analytical and numerical results is also presented in this chapter.

Keywords: elastic structure, moving loads, viscoelastic bearings, fractional-order, stochastic averaging method, Fokker-Planck-Kolmogorov equation

1. Introduction

There is a large amount of vehicles passing through in-service bridges every day, while sizable wind blows on the bridge decks. Vibrations caused by the service loads is of great theoretical and practical significance in civil engineering. In this chapter, it follows a list, by no means exhaustive, of research related to this kind engineer problem. To start with, Xu *et al.* [1] have explored the basic dynamics interaction between suspension bridges and the combined effects of intense wind and a single moving train; however, the interaction between the wind and the train dynamics has been altogether neglected. In this limit the suspension bridge response is dominated by wind force. The coupled dynamic analysis of vehicle and cable-stayed bridge system under turbulent wind has also been recently conducted by Xu and Guo [2] under the other limit of low wind speed. In the same view, the both effects of turbulent wind and moving loads on the brigde response are investigated numerically by Chen and Wu [3]. Another interesting results related to the problem of the dynamics of bridges subjected to the combined dynamic loads of vehicles and wind are presented in Refs. [4–6]. To summarize: from the standpoint of bridge engineering the disturbances, either due to wind (in low or high speed limits), passage of heavy loads (single massive trains or disordered aggregates of smaller freight carriers, result in a complex interaction with the bridge vibrations. However, the elastic properties of the bridges are enhanced by the insertion of bearings (the part ranging between the bridge deck and the piers) as a possible protection against severe earthquakes. For if one wants the bearings to protect the bridge, they should isolate the structure from ground vibrations and/or transfer the load to the foundation [7]. Noticed that the bearings can be constituted by some elastic or viscoelastic material. In the literature, the dynamics analysis of bridges with elastic bearings to moving loads has received limited attention. nevertheless, some authors like Yang *et al.* [8], Zhu and Law [9], Naguleswaran [10] and Abu Hilal and Zibdeh [11] have adressed a very interesting resultats about this subjet. There are investigated the pros, and the cons, of the elastic bearings.

The bearings can also be constituted by some viscoelastic materials (such as elastomer) [12]. Therefore, The viscoelastic property of the materials may be modelled by using the constitutive equation of Kelvin-Voigt type, which contain fractional derivatives of real order. In this Chapter we aim to investigate first the pros, and the cons, of the viscoelastic bearings and second the turbulence effect of the wind actions on the response of beam. To accomplish our goal some methods (analytical [13–15] and numerical [13–16]) are used.

2. Structural system model

2.1 Mathematical modelling

In this chapter, a simply supported Rayleigh beam [17, 18] of finite length L with geometric nonlinearities [19, 20], subjected by two kinds of moving loads (wind and train actions) and positioned on a foundation having fractional order viscoelastic physical properties is considered as structural system model and presented in **Figure 1**.

As demonstrated in (Appendix A) and in Refs. [16, 19, 21], the governing equation for small deformation of the beam-foundation system is given by:

$$\rho S\frac{\partial^2 w(x,t)}{\partial t^2}+EI\frac{\partial^4 w(x,t)}{\partial x^4}-\frac{3}{2}EI\frac{\partial^2}{\partial x^2}\left[\frac{\partial^2 w(x,t)}{\partial x^2}\left(\frac{\partial w(x,t)}{\partial x}\right)^2\right]-\rho I\frac{\partial^4 w(x,t)}{\partial x^2\partial t^2}+\mu\frac{\partial w(x,t)}{\partial t}$$

$$-\frac{ES}{2L}\frac{\partial^2 w(x,t)}{\partial x^2}\int_0^L\left(\frac{\partial w(x,t)}{\partial x}\right)^2\mathrm{d}x+\sum_{j=1}^{N_P}(k_j+c_jD_t^{\alpha_j})w(x,t)\delta\left[x-\frac{jL}{N_P+1}\right]=F_{ad}(x,t)+$$

$$P\sum_{i=0}^{N-1}\varepsilon_i\delta[x-x_i(t-t_i)]. \tag{1}$$

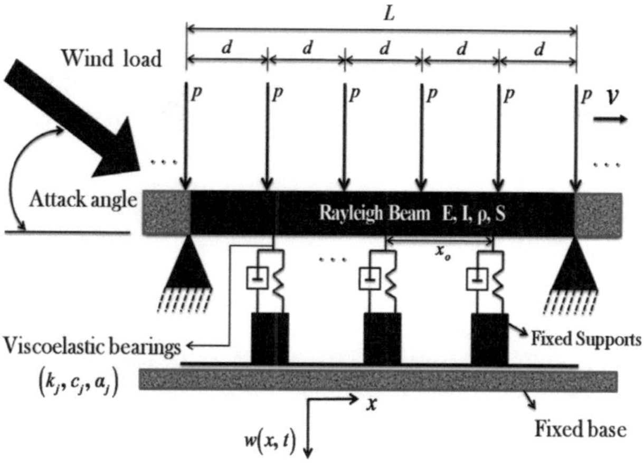

Figure 1.
Sketch of a beam-foundation system subjected to wind actions and series of moving forces.

In which ρS, EI, ρI, μ, $w(x,t)$, are the beam mass per unit length, the flexural rigidity of the beam, the transverse Rayleigh beam coefficient, the damping coefficient and the transverse displacement of the beam at point x and time t, respectively. In Eq. (1), $\rho S\frac{\partial^2 w(x,t)}{\partial t^2}$ represents the inertia term of the beam per unit length, $\rho I\frac{\partial^4 w(x,t)}{\partial x^2 \partial t^2}$ is the rotary inertia force of the beam per unit length, $\mu\frac{\partial w(x,t)}{\partial t}$ is the damping force of the beam per unit length, $\sum_{j=1}^{N_P}\left(k_j + c_j D_t^{\alpha_j}\right)w(x,t)\delta\left[x - \frac{jL}{N_P+1}\right]$ is the foundation-beam interaction force (per unit length of the beam's axis), $EI\frac{\partial^4 w(x,t)}{\partial x^4}$ and $\frac{3}{2}EI\frac{\partial^2}{\partial x^2}\left[\frac{\partial^2 w(x,t)}{\partial x^2}\left(\frac{\partial w(x,t)}{\partial x}\right)^2\right]$ are the linear and the nonlinear term of the rigidity of the beam essentially due to Euler Law. The nonlinear term is obtained by using the Taylor expansion of the exact formulation of the curvature up to the second order [19, 22, 23]. $\frac{ES}{2L}\frac{\partial^2 w(x,t)}{\partial x^2}\int_0^L\left(\frac{\partial w(x,t)}{\partial x}\right)^2 dx$ is the inplane tension of the beam [19, 20]. The terms on the right-hand side of Eq. (1) are used to describe the wind an train actions over the beam. In particular, the first term $F_{ad}(x,t)$ is the aerodynamic force given after some derivations by [24–26]:

$$F_{ad}(x,t) = \frac{1}{2}\rho_a b U^2\left[A_0 + \frac{A_1}{U}\frac{\partial w(x,t)}{\partial t} + \frac{A_2}{U^2}\left(\frac{\partial w(x,t)}{\partial t}\right)^2\right], \tag{2}$$

where ρ_a is the air mass density, b is the beam width, A_j (j = 0, 1, 2) are the aerodynamic coefficients (A_0 = 0.0297, A_1 = 0.9298, A_2 = -0.2400) [24]. U is the wind velocity which can be decomposed as $U = \overline{u} + u(t)$, where $\overline{u}$ is a constant (average) part representing the steady component and $u(t)$ is a time varying part representing the turbulence. It is assumed in this work that $(\overline{u} \gg u(t))$.

According to **Figure 1**, the boundary conditions of the beam are considered as [27]

$$w(0,t) = w(L,t) = 0; \quad \frac{\partial^2 w(0,t)}{\partial x^2} = \frac{\partial^2 w(L,t)}{\partial x^2} = 0. \tag{3}$$

The next section deals with the reduction of the main equation Eq. (1).

2.2 Reduced model equation

According to the Galerkin's method [27, 28] and by taking into account the boundary conditions of the beam, the solution of the partial differential Eq. (1) is given by:

$$w(x,t) = \sum_{n=1}^{\infty} q_n(t) \sin\left(\frac{n\pi x}{L}\right), \tag{4}$$

where $q_n(t)$ are the amplitudes of vibration and $\sin(n\pi x/L)$ are modal functions solutions of the beam linear natural equation with the associated boundary conditions. It is convenient to adopt the following dimensionless variables:

$$\chi_n = \frac{q_n}{l_r},\ \tau = \omega_0 t,\ \xi = \frac{u}{u_c}, \tag{5}$$

the single one-dimensional modal equation with $\chi_n = \chi(\tau)$ is given as:

$$\ddot{\chi}(\tau) + (2\lambda + \vartheta_1)\dot{\chi}(\tau) + \chi(\tau) + \beta\chi^3(\tau) + \eta\sum_{j=1}^{N_p}\left(k_j + c_j\omega_0^{\alpha_j}D_\tau^{\alpha_j}\right)\chi(\tau)\sin^2\left(\frac{j\pi}{N_p+1}\right)$$
$$= \vartheta_2\dot{\chi}^2(\tau) + \vartheta_0 + (\theta_0 + \theta_1\dot{\chi}(\tau))\xi(\tau) + f_0\sum_{i=0}^{N-1}\varepsilon_i \sin\Omega\left[\tau - i\frac{d\omega_0}{v}\right]. \tag{6}$$

With

$$\Omega = \frac{\pi v}{L\omega_0},\ f_0 = \frac{2PL^3}{l_r EI\pi^4},\ \eta = \frac{2L^3}{EI\pi^4},\ \beta = \frac{l_r^2}{4}\left[\frac{S}{I} - \frac{3}{2}\left(\frac{\pi}{L}\right)^2\right],$$
$$\vartheta_1 = \frac{\rho_a b L^3 A_1 \overline{U}}{2\pi^2\sqrt{EI\rho\left[L^2S + I\pi^2\right]}},\ \vartheta_0 = \frac{2\rho_a b A_0 \overline{U}^2 L^4}{EIl_r\pi^5},\ \vartheta_2 = \frac{4\rho_a b L^2 A_2 l_r}{3\rho\left[L^2S + I\pi^2\right]},$$
$$\theta_0 = \frac{2\overline{U}\rho_a b L^4 U_c A_0}{EIl_r\pi^6},\ \theta_1 = \frac{\rho_a b L^3 A_1 U_c}{2\pi^2\sqrt{EI\rho\left[L^2S + I\pi^2\right]}},\ \lambda = \frac{\mu L^3}{2\pi^2\sqrt{EI\rho\left[L^2S + I\pi^2\right]}}. \tag{7}$$

And

$$\omega_0 = \frac{\pi^2}{L}\sqrt{\frac{EI}{\rho(L^2S + I\pi^2)}},\ l_r = \frac{L}{2}. \tag{8}$$

According to Refs.[29, 30], Eq. (6) becomes:

$$\ddot{\chi}(\tau) + (2\lambda + \vartheta_1)\dot{\chi}(\tau) + \chi(\tau) + \beta\chi^3(\tau) + \eta\sum_{j=1}^{N_p}\left(k_j + c_j\omega_0^{\alpha_j}D_\tau^{\alpha_j}\right)\chi(\tau)\sin^2\left(\frac{j\pi}{N_p+1}\right)$$
$$= \vartheta_2\dot{\chi}^2(\tau) + \vartheta_0 + (\theta_0 + \theta_1\dot{\chi}(\tau))\xi(\tau) + F_{0N}\sin(\Omega\tau) - G_{0N}\cos(\Omega\tau). \tag{9}$$

Where

$$F_{ON} = P_0\left[1+\frac{2\sin\tilde{\tau}_0\sin((N-1)\tilde{\tau}_0)}{1-\cos(2\tilde{\tau}_0)}\cos(N\tilde{\tau}_0)\right],\ \tilde{\tau}_0=\frac{d\pi}{2L},$$
$$G_{ON} = \frac{2P_0\sin\tilde{\tau}_0\sin((N-1)\tilde{\tau}_0)}{1-\cos(2\tilde{\tau}_0)}\sin(N\tilde{\tau}_0). \tag{10}$$

3. Analytical explanation of the model

3.1 Effective analytical solution of the problem

In order to directly evaluate the response of the beam, the stochastic averaging method [13–15] is first applied to Eq. (6), then the following change in variables is introduced:

$$\chi(\tau) = a_0 + a(\tau)\cos\psi,\ \dot{\chi}(\tau) = -\Omega a(\tau)\sin\psi,\ \psi = \Omega\tau+\phi(\tau), \tag{11}$$

Substituting Eq. (11) into Eq. (9) we obtain:

$$\begin{cases} \dot{a}\cos\psi - a\dot{\varphi}\sin\psi = 0 \\ \dot{a}\sin\psi - a\dot{\varphi}\cos\psi = -\frac{1}{\Omega}[M_1(a,\psi)+M_2(a,\psi)]. \end{cases} \tag{12}$$

where

$$M_1(a,\psi) = F_{ON}\sin(\psi-\varphi) - G_{ON}\cos(\psi-\varphi) + (2\lambda+\vartheta_1)a\Omega\sin\psi a - \frac{1}{4}\beta a^3\cos 3\psi$$
$$-\left[1+\eta\sum_{j=1}^{N_p}k_j\sin^2\left(\frac{j\pi}{N_p+1}\right)+3\beta a_0^2+\frac{3}{4}\beta a^2-\Omega^2\right]\cos\psi+\frac{a^2}{2}\left[\vartheta_2\Omega^2-3\beta a_0\right]\cos 2\psi,$$
$$M_2(a,\psi) = -\eta\sum_{j=1}^{N_p}c_j\omega_0^{\alpha_j}\sin^2\left(\frac{j\pi}{N_p+1}\right)D_\tau^{\alpha_j}(a\cos\psi)+(\theta_0-\theta_1\Omega a\sin\psi)\xi(\tau). \tag{13}$$

According to Eq. (13) The derivatives of the generalized amplitude a and phase ϕ could be solved as:

$$\begin{cases} \dot{a} = -\frac{1}{\Omega}[M_1(a,\psi)+M_2(a,\psi)]\sin\psi \\ a\dot{\phi} = -\frac{1}{\Omega}[M_1(a,\psi)+M_2(a,\psi)]\cos\psi \end{cases} \tag{14}$$

a_0 satisfies the following non-linear equation:

$$\beta a_0^3 + \left[1+\frac{3}{2}\beta a^2+\eta\sum_{j=1}^{N_p}k_j\sin^2\left(\frac{j\pi}{N_p+1}\right)\right]a_0 = \vartheta_0 - \frac{1}{2}\vartheta_2\Omega^2 a^2. \tag{15}$$

Then, one could apply the stochastic averaging method [13–15] to Eq. (15) in time interval [0, T].

$$\begin{cases} \dot{a} = -\lim\limits_{T\to\infty} \dfrac{1}{T}\displaystyle\int_0^T \dfrac{1}{\Omega}[M_1(a,\psi) + M_2(a,\psi)]\sin\psi d\psi \\ a\dot{\varphi} = -\lim\limits_{T\to\infty} \dfrac{1}{T}\displaystyle\int_0^T \dfrac{1}{\Omega}[M_1(a,\psi) + M_2(a,\psi)]\cos\psi d\psi \end{cases} \tag{16}$$

According to this method, one could select the time terminal T as $T = 2\pi/\Omega$ in the case of periodic function ($M_1(a,\psi)$), or $T = \infty$ in the case of aperiodic one ($M_2(a,\psi)$). Accordingly, one could obtain the following pair of first order differential equations for the amplitude $a(\tau)$ and the phase $\phi(\tau)$:

$$\dot{a} = -(2\lambda - \vartheta_1)\frac{a}{2} - \frac{1}{2}\eta a \sum_{j=1}^{N_p} c_j \omega_0^{\alpha_j} \Omega^{\alpha_j - 1} \sin^2\left(\frac{j\pi}{N_p+1}\right)\sin\left(\frac{\alpha_j \pi}{2}\right) + \frac{1}{2\Omega}[G_{0N}\sin\varphi - F_{0N}\cos\varphi]$$

$$+\frac{\pi\theta_1^2 a}{8}[3S_\xi(2\Omega) + 2S_\xi(0)] + \frac{\pi\theta_0^2}{2\Omega^2 a}S_\xi(\Omega) + \sqrt{\frac{\pi\theta_0^2}{\Omega^2}S_\xi(\Omega) + \frac{\pi\theta_1^2 a^2}{4}[S_\xi(2\Omega) + 2S_\xi(0)]}\,\xi_1(\tau), \tag{17}$$

and

$$a\dot{\varphi} = \frac{a}{2\Omega}\left[1 + 3\beta a_0^2 - \Omega^2 + \frac{3}{4}\beta a^2 + \eta\sum_{j=1}^{N_p}\left(k_j + c_j\omega_0^{\alpha_j}\Omega^{\alpha_j}\cos\left(\frac{\alpha_j\pi}{2}\right)\right)\sin^2\left(\frac{j\pi}{N_p+1}\right)\right]$$

$$+\frac{1}{2\Omega}[G_{0N}\cos\varphi + F_{0N}\sin\varphi] - \frac{\pi\theta_1^2 a}{4}\Psi_\xi(2\Omega) + \sqrt{\frac{\pi\theta_0^2}{\Omega^2}S_\xi(\Omega) + \frac{\pi\theta_1^2 a^2}{4}S_\xi(2\Omega)}\;\xi_2(\tau). \tag{18}$$

Here $S_\xi(\Omega)$ and $\Psi_\xi(\Omega)$ are the cosine and sine power spectral density function, respectively [31]:

$$S_\xi(\varOmega) = \int_{-\infty}^{+\infty} R(\zeta)\cos(\varOmega\tau)\mathrm{d}\zeta = 2\int_0^{+\infty} R(\zeta)\cos(\varOmega\tau)\mathrm{d}\zeta = 2\int_{-\infty}^{0} R(\zeta)\cos(\varOmega\tau)\mathrm{d}\zeta,$$

$$\Psi_\xi(\varOmega) = 2\int_0^{+\infty} R(\zeta)\sin(\varOmega\tau)\mathrm{d}\zeta = -2\int_{-\infty}^{0} R(\zeta)\sin(\varOmega\tau)\mathrm{d}\zeta,$$

$$\int_{-\infty}^{+\infty} R(\zeta)\sin(\varOmega\tau)\mathrm{d}\zeta = 0;\;\; R(\zeta) = \mathrm{E}[\xi(\tau)\xi(\tau+\zeta)]. \tag{19}$$

In this work, $\xi(\tau)$ is assumed to be an harmonic function with constant amplitude σ_i and random phases $\gamma_i B_i(\tau) + \theta_i$. So, according to Refs. [31–33] the following model of $\xi(\tau)$ has been chosen:

$$\xi(\tau) = \sum_{i=1}^{m}\sigma_i\cos[\omega_i\tau + \gamma_i B_i(\tau) + \theta_i], \tag{20}$$

this model of the turbulent component of the wind $\xi(\tau)$ amounts to a bounded or cosine-Wiener noise, whose spectral density is given by:

$$\Phi_\xi(\omega) = \sum_{i=1}^{m} \frac{\sigma_i^2 \gamma_i^2 (\omega^2 + \omega_i^2 + \gamma_i^4/4)}{4\pi\left[(\omega^2 - \omega_i^2 - \gamma_i^4/4)^2 + \gamma_i^2\omega^2\right]}. \tag{21}$$

The next sections of this chapter will presents the analytical developments that we have made in order to express the beam response as a function of the system parameters. Then, let's start with the case where the beam is subjected to the moving loads only.

3.2 Analytical estimate of the beam response under moving loads only

We first consider system (1) with only deterministic moving loads ($F_{ad}(x,t) = 0$) neglecting wind effects on the beam. If $\vartheta_1 = \theta_0 = \theta_1 = 0$, Eqs. (15) and (16) become:

$$\begin{aligned} \dot{a} = &-\lambda a - \frac{1}{2}\eta a \sum_{j=1}^{N_p} c_j \omega_0^{\alpha_j} \Omega^{\alpha_j - 1} \sin^2\left(\frac{j\pi}{N_p+1}\right) \sin\left(\frac{\pi\alpha_j}{2}\right) \\ &+ \frac{1}{2\Omega}[G_{0N}\sin\phi - F_{0N}\cos\phi], \end{aligned} \tag{22}$$

and

$$\begin{aligned} a\dot{\varphi} = &\frac{a}{2\Omega}\left[1 - \Omega^2 + 3\beta a_0^2 + \frac{3}{4}\beta a^2 + \eta\sum_{j=1}^{N_p}\left(k_j + c_j\omega_0^{\alpha_j}\Omega^{\alpha_j}\cos\left(\frac{\pi\alpha_j}{2}\right)\right)\sin^2\left(\frac{j\pi}{N_p+1}\right)\right] \\ &+ \frac{1}{2\Omega}[G_{0N}\cos\varphi + F_{0N}\sin\varphi]. \end{aligned} \tag{23}$$

By substituting $a = A, \phi = \Phi$ and $\dot{a} = 0, \dot{\phi} = 0$ in Eqs. (22) and (23), algebraic manipulations give for the steady-state vibrations of the system response A the following non-linear equation:

$$\frac{9}{16}\beta^2 A^6 - \frac{3}{2}\beta\Theta_1(\alpha_j)A^4 + \left[\Theta_1^2(\alpha_j) + \Theta_2^2(\alpha_j)\right]A^2 = F_{0N}^2 + G_{0N}^2, \tag{24}$$

with

$$\begin{aligned} \Theta_1(\alpha_j) &= \Omega^2 - 1 - 3\beta a_0^2 - \eta\sum_{j=1}^{N_p}\left(k_j + c_j\omega_0^{\alpha_j}\Omega^{\alpha_j}\cos\left(\frac{\pi\alpha_j}{2}\right)\right)\sin^2\left(\frac{j\pi}{N_p+1}\right), \\ \Theta_2(\alpha_j) &= 2\Omega\lambda + \eta\sum_{j=1}^{N_p} c_j\omega_0^{\alpha_j}\Omega^{\alpha_j}\sin\left(\frac{\pi\alpha_j}{2}\right)\sin^2\left(\frac{j\pi}{N_p+1}\right). \end{aligned} \tag{25}$$

The stability of the steady-state vibration of the system response is investigated by using the method of Andronov and Witt [34] associated to the Routh-Hurwitz criterion [35]. Thus, the steady-state response is asymptotically stable if Eq. (26) is satisfied and unstable if Eq. (27) is satisfied:

$$\left(\frac{\Theta_2(\alpha_j)}{2\Omega}\right)^2 + \frac{1}{4\Omega}\left[\frac{3\beta}{4}A^2 - \Theta_1(\alpha_j)\right] \times \left[\frac{9\beta}{4}A^2 - \Theta_1(\alpha_j)\right] > 0, \tag{26}$$

$$\left(\frac{\Theta_2(\alpha_j)}{2\Omega}\right)^2 + \frac{1}{4\Omega}\left[\frac{3\beta}{4}A^2 - \Theta_1(\alpha_j)\right] \times \left[\frac{9\beta}{4}A^2 - \Theta_1(\alpha_j)\right] < 0. \tag{27}$$

The trivial solution of Eq. (15) is $a_0 = 0$.
What about the case where the beam is subjected to the stochastic wind loads?

3.3 Approximate solution of the beam response subjected to wind loads only

In this case $(F_{ad}(x,t) \neq 0)$ and $F_{0N} = G_{0N} = 0$, Eqs. (22) and (23) become:

$$da = \left\{-(2\lambda - \vartheta_1)\frac{a}{2} - \frac{1}{2}\eta a \sum_{j=1}^{N_p} c_j \omega_0^{\alpha_j} \Omega^{\alpha_j - 1} \sin^2\left(\frac{j\pi}{N_p+1}\right) \sin\left(\frac{\alpha_j \pi}{2}\right) + \frac{\pi\theta_1^2 a}{8}[3S_\xi(2\Omega) + 2S_\xi(0)]\right\} d\tau$$
$$+ \frac{\pi\theta_0^2}{2\Omega^2 a} S_\xi(\Omega) d\tau + \sqrt{\frac{\pi\theta_0^2}{\Omega^2} S_\xi(\Omega) + \frac{\pi\theta_1^2 a^2}{4}[S_\xi(2\Omega) + 2S_\xi(0)]} dW_1(\tau), \tag{28}$$

and

$$d\varphi = \frac{1}{2\Omega}\left\{1 + 3\beta a_0^2 - \Omega^2 + \frac{3}{4}\beta a^2 + \eta \sum_{j=1}^{N_p}\left(k_j + c_j \omega_0^{\alpha_j} \Omega^{\alpha_j} \cos\left(\frac{\alpha_j \pi}{2}\right)\right) \sin^2\left(\frac{j\pi}{N_p+1}\right)\right\} d\tau$$
$$- \frac{\pi\theta_1^2}{4}\Psi_\xi(2\Omega) d\tau + \sqrt{\frac{\pi\theta_0^2}{\Omega^2 a^2} S_\xi(\Omega) + \frac{\pi\theta_1^2 a^2}{4} S_\xi(2\Omega)}\, dW_2(\tau). \tag{29}$$

Here $W_1(\tau)$ and $W_2(\tau)$ are independent normalized Weiner processes. In order to evaluate the effects of wind parameters on the system response, we derive an evolution equation for the Probability Density Function (PDF) of the variable amplitude $a(\tau)$. The Fokker-Planck equation corresponding to the Langevin (Eq. (28)) reads:

$$\frac{\partial P(a,\tau)}{\partial \tau} = -\frac{\partial}{\partial a}\left[\left(-(2\lambda - \vartheta_1)\frac{a}{2} - \frac{1}{2}\eta a \sum_{j=1}^{N_p} c_j \omega_0^{\alpha_j} \Omega^{\alpha_j - 1} \sin^2\left(\frac{j\pi}{N_p+1}\right) \sin\left(\frac{\alpha_j \pi}{2}\right) + \frac{\pi\theta_0^2}{2\Omega^2 a} S_\xi(\Omega)\right) P(a,\tau)\right]$$
$$- \frac{\partial}{\partial a}\left[\frac{\pi\theta_1^2 a}{8}[3S_\xi(2\Omega) + 2S_\xi(0)] P(a,\tau)\right] + \frac{1}{2}\left(\frac{\pi\theta_0^2}{\Omega^2} S_\xi(\Omega) + \frac{\pi\theta_1^2 a^2}{4}[S_\xi(2\Omega) + 2S_\xi(0)]\right)\frac{\partial^2 P(a,\tau)}{\partial a^2}. \tag{30}$$

In the stationary case, $\frac{\partial P(a,\tau)}{\partial \tau} = 0$, the solution of Eq. (30) is:

$$P_s(a) = Na\left(\Gamma_0 + a^2 \Gamma_1\right)^{-(Q+1)}, \tag{31}$$

where

$$\Gamma_0 = \frac{\pi\theta_0^2}{\Omega^2} S_\xi(\Omega), \Gamma_1 = \frac{\pi\theta_1^2}{4}[S_\xi(2\Omega) + 2S_\xi(0)],\ Q = \frac{\Gamma_1 - 2\Gamma_2}{2\Gamma_1},$$

$$\Gamma_2 = -\frac{1}{2}(2\lambda - \vartheta_1) - \frac{1}{2}\eta \sum_{j=1}^{N_p} c_j \omega_0^{\alpha_j} \Omega^{\alpha_j - 1} \sin^2\left(\frac{j\pi}{N_p+1}\right) \sin\left(\frac{\pi\alpha_j}{2}\right) + \frac{\pi\theta_1^2}{8}[3S_\xi(2\Omega) + 2S_\xi(0)]. \tag{32}$$

Above N is a normalization constant that guarantees $\int_0^\infty P_s(a)\mathrm{d}a = 1$.

What about the case where the beam is subjected to the both moving vehicles and stochastic wind loads?

3.4 Approximate solution of the beam responses subjected to the both moving loads

Finally, the case where the beam is subjected to the series of lumped loads and the wind actions is investigated. For the analytical purposes, we assume that the beam is linear and it is submitted to only the additive effects of the wind loads. Thus, Eqs. (22) and (23) become:

$$a = \left[-(2\lambda - \vartheta_1)\frac{a}{2} - \frac{1}{2}\eta a \sum_{j=1}^{N_p} c_j \omega_0^{\alpha_j} \Omega^{\alpha_j - 1} \sin^2\left(\frac{j\pi}{N_p + 1}\right) \sin\left(\frac{\alpha_j \pi}{2}\right) + \frac{\Gamma_0}{a} \right] d\tau + \frac{1}{2\Omega}[G_{0N} \sin\varphi - F_{0N}\cos\varphi] d\tau + \sqrt{\Gamma_0} dW_1(\tau), \tag{33}$$

and

$$d\varphi = \frac{1}{2\Omega}\left[1 - \Omega^2 + \eta \sum_{j=1}^{N_p} \left(k_j + c_j \omega_0^{\alpha_j} \Omega^{\alpha_j} \cos\left(\frac{\alpha_j \pi}{2}\right)\right) \sin^2\left(\frac{j\pi}{N_p+1}\right)\right] d\tau + \frac{1}{2\Omega a}[G_{0N}\cos\varphi + F_{0N}\sin\varphi] d\tau + \frac{1}{a}\sqrt{\Gamma_0} dW_2(\tau). \tag{34}$$

The averaged Fokker-Planck-Kolmogorov equation associated with the previous Itô Eqs. (33) and (34) is

$$\frac{\partial P(a,\phi,\tau)}{\partial \tau} = -\frac{\partial}{\partial a}(\overline{a}_1 P(a,\phi,\tau)) - \frac{\partial}{\partial \phi}(\overline{a}_2 P(a,\phi,\tau)) + \frac{1}{2}\frac{\partial^2}{\partial a^2}(\overline{b}_{11} P(a,\phi,\tau)) + \frac{1}{2}\frac{\partial^2}{\partial \phi^2}(\overline{b}_{22} P(a,\phi,\tau)), \tag{35}$$

where

$$\begin{aligned}
\overline{a}_1 &= -(2\lambda - \vartheta_1)\frac{a}{2} - \frac{1}{2}\eta a \sum_{j=1}^{N_p} c_j \omega_0^{\alpha_j} \varOmega^{\alpha_j - 1} \sin^2\left(\frac{j\pi}{N_p+1}\right) \sin\left(\frac{\alpha_j\pi}{2}\right) + \frac{1}{2\Omega}[G_{0N}\sin\phi - F_{0N}\cos\phi] + \frac{\Gamma_0}{a} \\
\overline{a}_2 &= \frac{1}{2\Omega}\left[1 - \varOmega^2 + \eta \sum_{j=1}^{N_p}\left(k_j + c_j \omega_0^{\alpha_j} \varOmega^{\alpha_j} \cos\left(\frac{\alpha_j \pi}{2}\right)\right) \sin^2\left(\frac{j\pi}{N_p+1}\right) + \frac{1}{a}[G_{0N}\cos\phi + F_{0N}\sin\phi] \right] \\
\overline{b}_{11} &= \varGamma_0 \\
\overline{b}_{22} &= \frac{\varGamma_0}{a^2}.
\end{aligned} \tag{36}$$

Applying the solution procedure proposed by Huang *et al.* [36], one obtains the following exact stationary solution

$$P_s(a,\phi) = N' a \exp\left\{ \frac{\varGamma_2'}{\varGamma_0} a^2 - \frac{a}{\Omega(\Gamma_0^2 + d_0^2)}[(d_0 G_{0N} + F_{0N}\varGamma_0)\cos\phi + (d_0 F_{0N} - G_{0N}\varGamma_0)\sin\phi] \right\} \tag{37}$$

where N' is a normalization constant and

$$\Gamma_2' = -\frac{1}{2}(2\lambda - \vartheta_1) - \frac{\eta}{2}\sum_{j=1}^{N_p} c_j \omega_0^{\alpha_j} \Omega^{\alpha_j - 1} \sin^2\left(\frac{j\pi}{N_p+1}\right) \sin\left(\frac{\alpha_j \pi}{2}\right), d_0 = -\frac{\Gamma_0 \Gamma_3}{\Gamma_2'},$$

$$\Gamma_3 = \frac{1}{2\Omega}\left[1 - \Omega^2 + \eta \sum_{j=1}^{N_p}\left(k_j + c_j \omega_0^{\alpha_j} \Omega^{\alpha_j} \cos\left(\frac{\alpha_j \pi}{2}\right)\right) \sin^2\left(\frac{j\pi}{N_p+1}\right)\right]. \tag{38}$$

4. Numerical analysis of the model

All the parameters concerning the chosen models of beam, of foundation and of the aerodynamic force are presented in Ref. [21]. Theses parameters clearly help to calculated the dimensionless coefficients defined in Eq. (7). It is well know that the validation of the results obtained through analytical investigation is guarantee by the perfect match with the results obtained through numerical simulations. thus, The numerical scheme used in this chapter is based on the Grunwald-Letnikov definition of the fractional order derivative Eq. (39). [37–40] and the Newton-Leipnik algorithm [37, 38]

$$D_\tau^\alpha\left[\chi\left(\tau_{n_f}\right)\right] \approx h^{-\alpha} \sum_{l=0}^{n_f} C_l^\alpha \chi\left(\tau_{n_f - l}\right), \tag{39}$$

where h is the integration step and the coefficients C_l^α satisfy the following recursive relations:

$$C_0^\alpha = 1, \quad C_l^\alpha = \left(1 - \frac{1+\alpha}{l}\right) C_{l-1}^\alpha. \tag{40}$$

Now we display in some figures the effects of the main parameters of the proposed model. For example, **Figure 2(a)** shows the effect of the number of the bearings on the amplitude of vibration of the beam. This graph also shows a

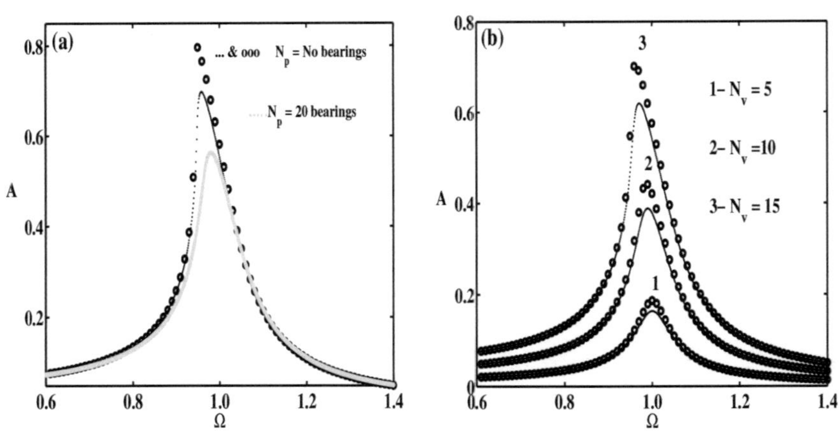

Figure 2.
Effects of the number of the bearings N_p (a) and number of moving loads N_v (b) on the amplitude response of the beam when driving frequency Ω: Analytical curves (-), numerical curves (o). All The parameters are given in [21].

comparison between the results from the mathematical analysis (curve with dotted line) and the results obtained from numerical simulation of Eq. (9) using Eq. (39). The match between the results shows a good level of precision of the approximation made in obtaining Eq. (24). This figure also reveals that the vibration amplitude of the beam decreases and the resonance frequency of the system increases as the number of bearings increases. **Figure 2(b)**, shows the effect of loads number on the beam response. It is observed that as the value of N_v increases, the amplitude of vibration at the resonant state merely increases.

Looking at the effects of the order of the fractional derivative α on the amplitude of the beam, we obtain the graph of **Figure 3**. Small value of α leads to large value of the maximum vibration amplitude. It is also clearly shown that the system is more stable for the highest order of the derivative. The multivalued solution appears for small order and disappears progressively as the order increases. The resonance (a peak of the amplitude) appears as the parameter k_0' increases, see **Figure 3(a)-(d)**. The good match between the analytical and the numerical results gives a validation of the approximations made.

Also, the stochastic analysis has allowed to estimate the probabilistic distribution as a consequence of the wind random effects. The beam response, and more specifically the stationary probability density function $P_s(a)$ of its amplitude a, can also be retrieved (**Figures 4** and **5**). This type of analysis indicates that as the additive wind turbulence parameter increases, the peak value of the probability density function

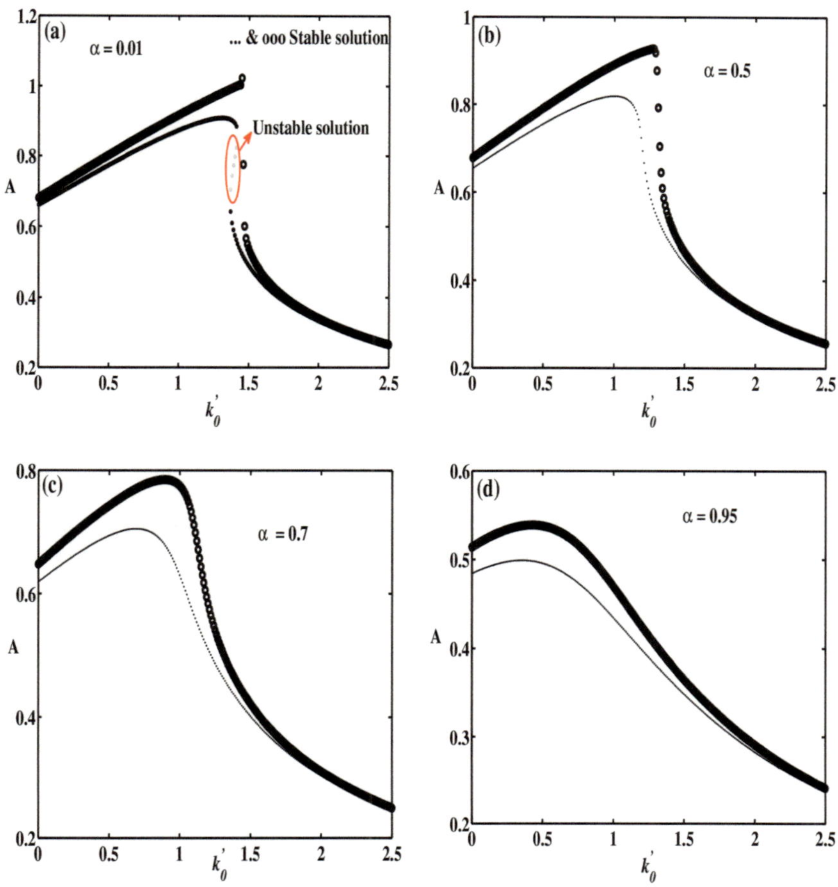

Figure 3.
Effect of the fractional order on the amplitude of the beam A when the dimensionless stiffness k_0' varying: Analytical curves (-), numerical curve (o). All The parameters are given in [21].

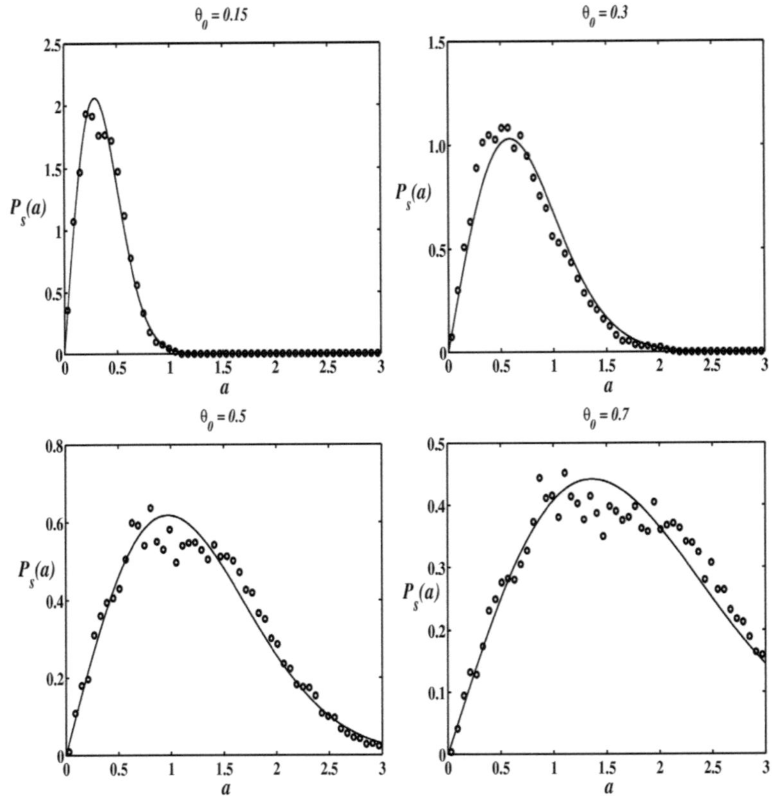

Figure 4.
Stationary probability distribution function as function of the amplitude for different values of the additive wind turbulence parameter. Analytical curves (-), numerical curve (o). All The parameters are given in [21].

decreases and progressively shifts toward larger amplitude values, while the average center position stays in the same position. Thus, the additive (θ_0) and parametric (θ_1) wind turbulence decreases the chance for the beam to quickly reach the amplitude resonance. It is also demonstrated that the PDF has only one maximum situated in the vicinity of $a_m = 0.2$.

We have plotted curves **Figure 6(a)** and **(b)** that presents the stationary probability distribution function $P_s(a, \phi)$ versus the amplitude a and the phase ϕ. This graph just confirm the results obtained in **Figures 4** and **5** and the highest value of the PDF is more visible.

Figure 7 presents the times histories of the maximum vibration of the beam. The case where the beam is subjected to moving loads (a), to wind actions (b) and to the both wind and loads (c).

5. Conclusions

In this Chapter we have revised some aspects of the response of viscoelastic foundation of bridges to a simplified model of moving loads and wind random perturbations. The results have been compared to the numerical solution for the modal equation, obtained with the deterministic and stochastic version of Newton-Leipnik algorithm. The analysis has begun modeling the steady-state vibration of the beam suspensions made of a fractional-order viscoelastic material. The resulting mathematical model consists of a component for the beam, and the Kelvin-Voigt foundation type containing fractional derivative of real order, as well as a stochastic

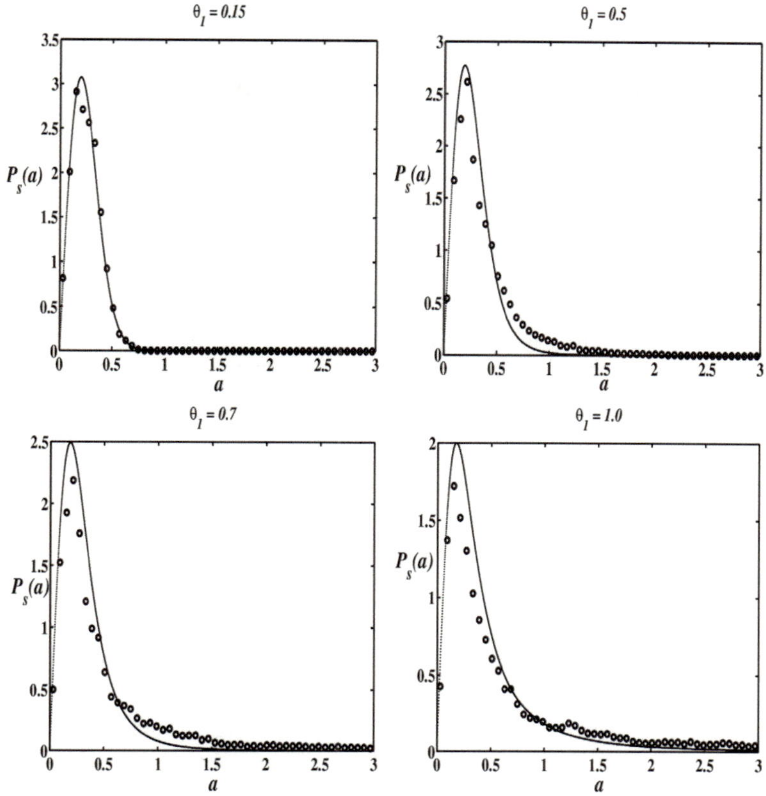

Figure 5.
Stationary probability distribution function as function of the amplitude for different values of the parametric wind turbulence parameter. Analytical curves (-), numerical curve (o). All The parameters are given in [21].

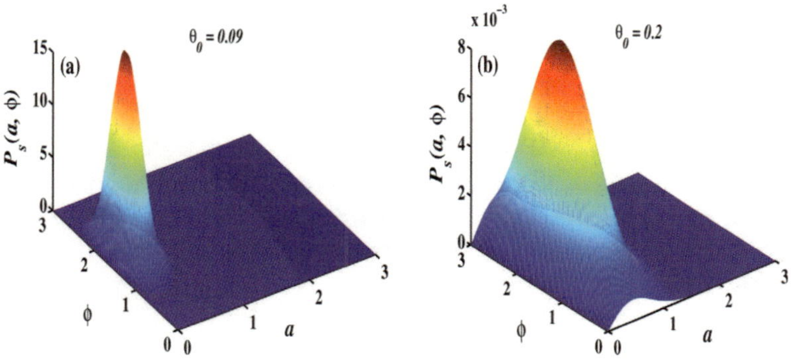

Figure 6.
Stationary probability distribution function versus the amplitude a and the phase ϕ for (a) $\theta_0 = 0.09$ and (b) $\theta_0 = 0.2$. All The parameters are given in [21].

term to account for wind pressure. We have highlighted the simplifications. Perhaps the most significant, in the very model formulation, has been the assumption that the load passage consists of concentrated masses, spatially periodic and moving at constant speed. This simplification is crucial to reduce the full partial differential equation to a single smode model — Wind load is modeled as the aerodynamic force related to the wind that blows orthogonally to the beam axis with random velocity. The whole system has then been modeled with a partial differential equation that can be reduced to a one-dimensional modal equation. The beam response under moving and/or stochastic wind loads has been estimated analytically assuming that

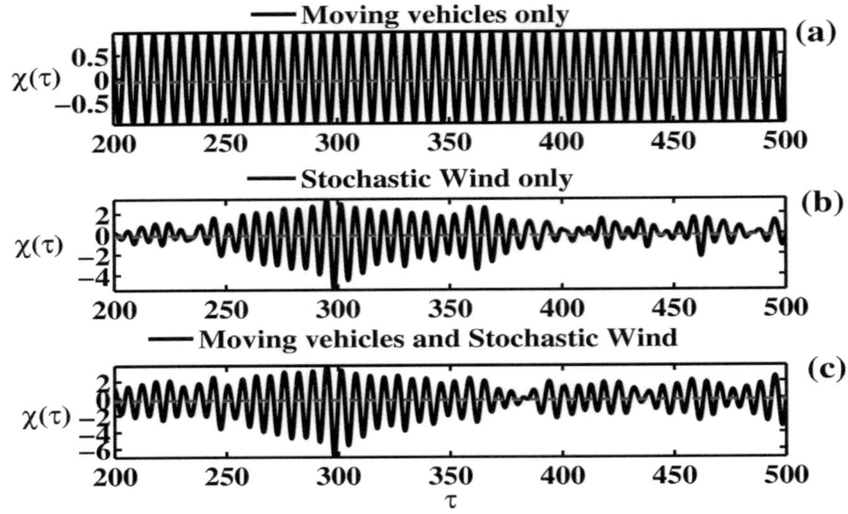

Figure 7.
Vibration amplitude of the beam $\chi(\tau)$ as function of the time τ. All The parameters are given in [21].

the first mode contains the essential information, and using the stochastic averaging method. The analysis has therefore some limitations, namely the limited values of parameters that have been explored, and to have retained the first mode only in the Galerkin's method. Also, the vehicles' train has been (over?) simplified as a simple periodic drive. With this limitations in mind, let us summarize the main findings. To start with, in this framework it is possible to investigate how the main parameters of the moving loads and of the bearings affect the beam response, and especially how the driving frequency, the loads number, the stiffness coefficient, fractional-order of the viscosity term and the number of bearings affect the dynamic behaviour of the beam. The resonance phenomenon and the stability in the beam system strongly depends on the stiffness and fractional-order of the derivative term of the viscous properties of the bearings. There are a number of quantitative results that are worth mentioning. Firstly, as the number of moving loads increases, the resonant amplitude of the beam increases as well. Secondly, it has been established that as the number of bearings increases, the resonant amplitude decreases and, more importantly, shifts toward larger frequency values. Thirdly, the system response becomes more stable as the order of the derivative increases, for the multivalued solution only appears for the smallest order and quickly disappears as the order increases. All the above results are a consequence of the analysis of the oscillations. However, the stochastic analysis has allowed to estimate the probabilistic distribution as a consequence of the wind random effects. The beam response, and more specifically the stationary probability density function of its amplitude, can also be retrieved. This type of analysis indicates that as the additive wind turbulence parameter increases, the peak value of the probability density function decreases and progressively shifts toward larger amplitude values, while the average center position stays in the same position. Thus, the additive and parametric wind turbulence decreases the chance for the beam to quickly reach the amplitude resonance.

Numerical simulations have confirmed these predictions. This behavior is depicted in **Figures 2–5**, which have practical implications, on which we would like to comment. To make an example, the beam system frequency Ω displays the bridge response as the vehicles speed changes, see Eq. (7). In principle it could be possible to avoid large oscillations by controlling the speed of the freight vehicles, albeit in practice it is more realistic to set a maximum speed at which the bridge should be crossed. In other words, to keep resonance at bay, it is necessary to set a speed limit below the resonance insurgence. Analogously, one could think to limit the vehicles number, not to have a minimum number of vehicles across the bridge. We conclude the parameters analysis, namely the stiffness and the viscoelastic properties of the foundations, noticing that such parameters can be optimized with an appropriated tuning, see e.g., **Figure 3**, or the analogous indications that stems from the results of **Figures 4** and **5** for the wind features θ_0 and θ_1.

By way of conclusion, let us summarize that the special properties of the visco-elastic foundations and of the time dependent perturbations, vehicles and wind, interact. As a result also the construction and management parameters are not to be considered independent procedures, for they are deeply interwoven if safe transportation is to be guaranteed.

Acknowledgements

Part of this work was completed during a research visit of Prof. Nana Nbendjo at the University of Kassel in Germany. He is grateful to the Alexander von Humboldt Foundation for financial support within the Georg Forster Fellowship.

Appendix A

To deal with the modelling, let us consider the dynamic equilibrium of a beam element of length dx; $w = w(x,t)$ and $\theta = \theta(x,t)$ be the transversal displacement and the angle of rotation of the beam element respectively. We denote the internal bending moment by M, the internal shear force by V, the inplane tension due to the inplane strain, issue of the assumed negligible longitudinal displacement of the beam by T, the foundation-beam interaction force (per unit length of the beam's axis) by $Q_F(x,t)$ and the external distributed loading by $F_{ad}(x,t)$ and $f(x,t)$.

Setting the vertical forces on the element equal to the mass times acceleration gives:

$$\frac{\partial V}{\partial x} = Q_F(x,t) - f(x,t) - F_{ad}(x,t) + \rho S \frac{\partial^2 w(x,t)}{\partial t^2} \tag{41}$$

While summing moments produces:

$$\frac{\partial M}{\partial x} = V - \rho I \frac{\partial^2 \theta(x,t)}{\partial t^2} - T \frac{\partial w(x,t)}{\partial x} \tag{42}$$

For small rotation $\theta(x,t) \approx \frac{\partial w(x,t)}{\partial x}$, Eq. (42) becomes:

$$\frac{\partial M}{\partial x} = V - \rho I \frac{\partial^3 w(x,t)}{\partial t^2 \partial x} - T \frac{\partial w(x,t)}{\partial x} \tag{43}$$

Combining Eq. (41) and Eq. (43) then yields:

$$\frac{\partial^2 M}{\partial x^2} = Q_F(x,t) - f(x,t) - F_{ad}(x,t) + \rho S \frac{\partial^2 w(x,t)}{\partial t^2} - \rho I \frac{\partial^4 w(x,t)}{\partial t^2 \partial x^2} - T \frac{\partial^2 w(x,t)}{\partial x^2} \tag{44}$$

From the geometry of the deformation, and using Hooke's law $\sigma_x = E\varepsilon_x$, one can show that (see reference [19]):

$$M = \frac{EI}{R} = -EI\frac{\frac{\partial^2 w(x,t)}{\partial x^2}}{\left[1+\left(\frac{\partial w(x,t)}{\partial x}\right)^2\right]^{\frac{3}{2}}} \approx -EI\frac{\partial^2 w(x,t)}{\partial x^2}\left[1-\frac{3}{2}\left(\frac{\partial w(x,t)}{\partial x}\right)^2\right] + O\left(\left(\frac{\partial w(x,t)}{\partial x}\right)^2\right)$$

$$\approx -EI\frac{\partial^4 w(x,t)}{\partial x^4} + \frac{3}{2}EI\frac{\partial^2}{\partial x^2}\left[\frac{\partial^2 w(x,t)}{\partial x^2}\left(\frac{\partial w(x,t)}{\partial x}\right)^2\right] + O\left(\left(\frac{\partial w}{\partial x}\right)^2\right) \tag{45}$$

where the Taylor expansion of the inverse of the radius of curvature $(\frac{1}{R})$ up to the second order is carried out. According to the assumed negligible longitudinal displacement of the beam, the tension in the beam T can be determined as (see the details of their derivation in Ref.[19]).

$$T = \frac{ES}{2L}\int_0^L \left(\frac{\partial w(x,t)}{\partial x}\right)^2 \mathrm{d}x \tag{46}$$

Finally taking into account the dissipation $(\mu\frac{\partial w(x,t)}{\partial t})$, putting Eq. (44), Eq. (45) and Eq. (46) together gives the new desired result (Eq. (1) of the manuscript)

$$\rho S\frac{\partial^2 w(x,t)}{\partial t^2} - \rho I\frac{\partial^4 w(x,t)}{\partial x^2 \partial t^2} + EI\frac{\partial^4 w(x,t)}{\partial x^4} + \mu\frac{\partial w(x,t)}{\partial t} - \frac{3}{2}EI\frac{\partial^2}{\partial x^2}\left[\frac{\partial^2 w(x,t)}{\partial x^2}\left(\frac{\partial w(x,t)}{\partial x}\right)^2\right],$$

$$-\frac{ES}{2L}\frac{\partial^2 w(x,t)}{\partial x^2}\int_0^L \left(\frac{\partial w(x,t)}{\partial x}\right)^2 \mathrm{d}x + Q_F(x,t) = F_{ad}(x,t) + f(x,t) \tag{47}$$

where:

$$f(x,t) = P\sum_{i=0}^{N-1} \varepsilon_i \delta[x - x_i(t-t_i)]$$

$$Q_F(x,t) = \sum_{j=1}^{N_P}(k_j + c_j D_t^{\alpha_j})w(x,t)\delta\left[x - \frac{jL}{N_P+1}\right]$$

Author details

Lionel Merveil Anague Tabejieu[1*], Blaise Roméo Nana Nbendjo[2] and Giovanni Filatrella[3]

1 Department of Mechanical Engineering, National Higher Polytechnic School of Douala, University of Douala, Douala, Cameroon

2 Laboratory of Modelling and Simulation in Engineering, Biomimetics and Prototypes, Faculty of Science, University of Yaounde I, Yaounde, Cameroon

3 University of Sannio, Via Port' Arsa 11, Benevento, Sannio, Italy

*Address all correspondence to: lmanaguetabejieu@gmail.com

References

[1] Xu YL, Xia H and Yan QS. Dynamic response of suspension bridge to high wind and running train. Journal of Bridge Engineering. 1988; 8: 46-55.

[2] Xu YL and Guo WH. Dynamic analysis of coupled road vehicle and cable-stayed bridge system under turbulent wind. Engineering Structures. 2003; 25: 473-486.

[3] Chen SR and Wu J. Dynamic performance simulation of long-span bridge under combined loads of stochastic traffic and wind. Journal of Bridge Engineering. 2010; 3: 219-230.

[4] Zhou y and Chen S. Dynamic Simulation of a Long-Span Bridge-Traffic System Subjected to Combined Service and Extreme Loads. Journal of Structural Engineering. 2014;04014215.

[5] Zhang W, Cai CS, Pan F. Fatigue Reliability Assessment for Long-Span Bridges under Combined Dynamic Loads from Winds and Vehicles. Journal of Bridge Engineering. 2013; 8: 735-747.

[6] Li Y, Qiang S, Liao H, Xu YL. Dynamics of wind-rail vehicle-bridge systems. Journal of Wind Engineering and Industrial Aerodynamics. 2005; 93: 483-507.

[7] Yau JD, Wu YS and Yang YB. Impact response of bridges with elastic bearings to moving loads. Journal of Sound and Vibration. 2001; 248: 9-30.

[8] Yang YB, Lin CL, Yau JD and Chang DW. Mechanism of resonance and cancellation for train-induced vibrations on bridges with elastic bearings. Journal of Sound and Vibration. 2004; 269: 345-360.

[9] Zhu XQ, Law SS. Moving load identification on multi-span continuous bridges with elastic bearings. Mechanical Systems and Signal Processing. 2006; 20; 1759-1782.

[10] Naguleswaran S. Transverse vibration of an Euler-Bernoulli uniform beam on up to five resilient supports including ends. Journal of Sound and Vibration. 2003; 261: 372-384.

[11] Abu Hilal M, Zibdeh HS. Vibration analysis of beams with general boundary conditions traversed by a moving force. Journal of Sound and Vibration. 2000; 229: 377-388.

[12] Barone G, Di Paola M, Lo Iacono F, Navarra G. Viscoelastic bearings with fractional constitutive law for fractional tuned mass dampers. Journal of Sound and Vibration. 2015; 344: 18-27.

[13] Stratonovich RL. Selected Topics in the Theory of random Noise. Gordon and Breach, New York vols.1 and 2, 1963.

[14] Roberts JB, Spanos PD. Stochastic averaging: An approximate method of solving random vibration problems. International Journal of Non-Linear Mechanic. 1986; 21: 111-134.

[15] Vladimir B. Method of Averaging for Differential Equations on an Infinite Interval: Theory and Applications. Taylor & Francis Group, New York, 2007.

[16] Anague Tabejieu LM, Nana Nbendjo BR, Woafo P. On the dynamics of Rayleigh Beams resting on Fractional-order Viscoelastic Pasternak Foundations subjected to moving loads. Chaos Solitons and Fractals. 2016; 93: 39-47.

[17] Anague Tabejieu LM, Nana Nbendjo BR, Filatrella G, Woafo P. Amplitude stochastic response of Rayleigh beams to randomly moving loads. Nonlinear Dynamics. 2017; 89: 925-937.

[18] Han SM, Benaroya H, Wei T. Dynamics of transversely vibrating

beams using four engineering theories. Journal of Sound and Vibration. 1999; 225: 935-988.

[19] Kovacic I and Brennan MJ. The Duffing Equation: Nonlinear Oscillators and their Behaviour. Wiley, New york, 2011.

[20] Nayfeh AH, Mook DT. Non Linear Oscillations. Wiley-Interscience, New york, 1979.

[21] Anague Tabejieu LM, Nana Nbendjo BR, Filatrella G. Effect of the fractional foundation on the response of beam structure submitted to moving and wind loads. Chaos, Solitons and Fractals. 2019; 127: 178-188.

[22] Anague Tabejieu LM, Nana Nbendjo BR, Dorka U. Identification of horseshoes chaos in a cable-stayed bridge subjected to randomly moving loads. International Journal of Non-linear Mechanics. 2016; 85: 62-69.

[23] Oumbé Tékam GT, Tchawou Tchuisseu EB, Kitio Kwuimy CA, Woafo P. Analysis of an electromechanical energy harvester system with geometric and ferroresonant nonlinearities. Nonlinear Dynamics. 2014; 76: 1561-1568.

[24] Luongo A, Zulli D. Parametric, external and self-excitation of a tower under turbulent wind flow. Journal of Sound and Vibration. 2011, 330: 3057-3069.

[25] Simiu E and Scanlan RH. Wind effects on structures, fundamentals and applications to design. John Wiley & Sons, New york, (1996).

[26] Van Horssen WT. An asymptotic theory for a class of initial-boundary value problems for weakly nonlinear wave equations with an application to a model of the galloping oscillations of overhead transmission lines. SIAM Journal on Applied Mathematics. 1988; 48: 1227-1243.

[27] Timoshenko S and Gere JM. Theory of Elastic Stability. McGraw-Hill, New York, 1961.

[28] Kitio Kwuimy CA, Nana Nbendjo BR, Woafo P. Optimization of electromechanical control of beam dynamics: Analytical method and finite differences simulation. Journal of Sound and Vibration 2006; 296: 180-193.

[29] Nikkhoo A, Hassanabadi ME, Azam SE, Amiri JV. Vibration of a thin rectangular plate subjected to series of moving inertial loads. Mechanics Research Communications. 2014; 55: 105-113.

[30] Mangulis V. Handbook of Series for Scientists and Engineers. Academic Press, New York, 1965.

[31] Xie W-C. Dynamic stability of structures. Cambridge university press, New york, 2006.

[32] Lin YK and Li QC. New stochastic theory for bridge stability in turbulence flow. Journal of Engineering Mechanics. 1993; 119: 113-127.

[33] Cai GQ, Wu C. Modeling of bounded stochastic processes. Probabilistic Engineering Mechanics. 2004; 19: 197-203.

[34] Andronov AA, Witt A. Towards mathematical theory of capture. Archiv fur Electrotechnik. 1930; 24: 99-110.

[35] Awrejcewicz J., Krysko VA. Introduction to Asymptotic Methods. CRC Press, New York, 2006.

[36] Huang ZL and Zhu WP. Exact stationary solutions of averaged equations of stochastically and harmonically excited MDOF quasi-linear systems with internal and or external resonances. Journal of Sound and Vibration. 1997; 204: 249-258.

[37] Debnath L. Recent applications of fractional calculus to science and

engineering. International Journal of Mathematics and Mathematical Science. 2003; 54: 3413-3442.

[38] Petráš I. Fractional-order Nonlinear Systems: Modeling, Analysis and Simulation. Higher Education Press, Beijing, 2011.

[39] Podlubny I. Fractional Differential Equations. Academic Press, San Diego: USA, 1999.

[40] Ortigueira MD. Fractional Calculus for Scientists and Engineers: Lecture Notes in Electrical Engineering. Springer, Berlin, 2011.

Chapter 3

Chaotic Dynamics and Complexity in Real and Physical Systems

Mrinal Kanti Das and Lal Mohan Saha

Abstract

Emergence of chaos and complex behavior in real and physical systems has been discussed within the framework of nonlinear dynamical systems. The problems investigated include complexity of Child's swing dynamics , chaotic neuronal dynamics (FHN model), complex Food-web dynamics, Financial model (involving interest rate, investment demand and price index) etc. Proper numerical simulations have been carried out to unravel the complex dynamics of these systems and significant results obtained are displayed through tables and various plots like bifurcations, attractors, Lyapunov exponents, topological entropies, correlation dimensions, recurrence plots etc. The significance of artificial neural network (ANN) framework for time series generation of some dynamical system is suggested.

Keywords: chaos, Poincarè map, bifurcation, Lyapunov exponents, topological entropy, correlation dimension, permutation entropy, neural network

1. Introduction

In this chapter, we investigate the dynamical complexity of several real physical systems. We present our analysis of various problems considered here and present results graphically based on actual numerical simulation for various system. We revisit the analysis of complexity of nonlinear pendulum dynamics and its application to unravel the complex oscillations observed in a swing pumped by a child. For the analysis, we use various tools e.g., phase plot, bifurcation diagram, Poincare surface of section and maps, Lyapunov exponent (LCE) etc., of theory of nonlinear dynamical system. Next we consider the problem of prey-predator system with Allee effect and introduce correlation dimension and topological entropy to characterize the fractal structure and the associated complexity in its dynamics. Further, beside the normal analysis used to understand the complex neuronal dynamics, say using Fitzhug-Nagumo model (FHN), recurrence plots (RPs) have been used along with the phase plot analysis and bifurcation diagram to picturise the transition of spike occurrence from periodic to quasi-periodic and chaotic oscillations in the presence of external periodic stimulation. Significance of multi-scale permutation entropy analysis to characterize nonlinear dynamical complexity of real system is also suggested while analyzing a financial system involving interest rate, investment demand and interest rates. Finally, we describe the utility of time series generation of dynamical variables of chaotic system, such as Lorentz system, using artificial neural network.

2. Pendulum motion and dynamics behind a swing

2.1 Pendulum motion

Motion of nonlinear driven pendulum with friction are widely discussed through numerous literature in Physics and applied Mathematics, (e.g. here, [1–3]). The nonlinear analysis of driven nonlinear pendulum provides a basis for understanding the complexity of various nonlinear dynamical systems. Regular and chaotic motions are observed in such pendulums depending on the numerical values assigned to the parameters associated in their equations of motion. A swing dynamics is very similar to that of nonlinear driven pendulum, [4–7]. In the present text regular and chaotic motion of a pendulum and that of the child's swing is discussed mathematically. Numerical results are presented in various forms of graphics. The equation of motion of a driven pendulum having angular displacement, θ, from vertical with linear damping expressed as

$$\frac{d^2\theta}{dt^2} + k\frac{d\theta}{dt} + \omega_0^2 \sin\theta = F\cos(\omega t) \tag{1}$$

where F and ω are respectively the amplitude and frequency of the driving force and k is the damping coefficient and $\omega_0 = \sqrt{\frac{g}{L}}$ is the natural frequency for free small-amplitude oscillations. Here g refers to acceleration due to gravity and L the length of the pendulum. Often it is convenient to express frequency in units of ω_0 by setting $\omega_0 \to 1$ and rescaling the time unit accordingly. The periodic force $F\cos\omega t$ is active and influence the motion of the pendulum.

The Eq. (1) can easily be replaced by equation

$$\frac{d^2\theta}{dt^2} + 2\beta\frac{d\theta}{dt} + \omega_0^2 \sin\theta = f\omega_0^2 \cos(\omega t) \tag{2}$$

Here $k = 2\beta$, $\beta = \frac{\gamma}{m}$, i.e., ratio of damping coefficient per unit mass m, $\omega_0^2 = \frac{g}{L}$ and $f = \frac{F}{\omega_0^2}$. One obtains a bifurcation diagram for Eq. (2), shown in **Figure 1**. Bifurcation scenario indicates a period doubling phenomena followed by chaos. This implies the pendulum oscillations may be regular or chaotic depending on the magnitudes of external forcing.

System (1) or (2) are very common structure with very few degrees of freedom. The simple forced pendulum is periodic in θ when the driving force applied is

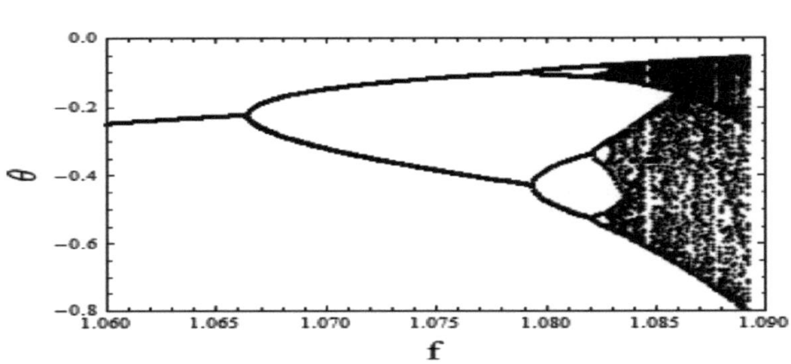

Figure 1.
Bifurcation scenario of damped and driven pendulum for values $\beta = 0.75$, $\omega = 2\pi$, $\omega_o = 3\pi$ and $1.06 \leq f \leq 1.09$.

smaller than a certain threshold f_C and chaotic when the force f is greater than this threshold. An interesting case would also be when the driving force becomes comparable to the weight.

Keeping fixed values for $k = 0.125$, $\omega_0 = 1$, $\omega = 2/3$ and then varying forcing amplitude F, following regular (periodic) and chaotic motion of the pendulum is observed:

Case (a): for a value $F = 0.2$, a time-series plot, phase plot, surface of section and Poincaré map, are drawn as shown in **Figure 2**.

Case (b): for a value $F = 0.8$, corresponding figures of case (a) are obtained as a time-series plot, phase plot, surface of section and Poincaré map, are drawn as shown in **Figure 3**.

The plots shown in **Figure 3** indicate at a value $F = 0.8$ the pendulum oscillation is chaotic and this leads to unpredictability. Pendulum may be whirling irregularly or overturn or show very irregular oscillations.

As an application of the foregoing analysis, in the following section, we extend the formalism to discuss the problem of swing oscillation where the length of the pendulum varies periodically.

2.2 Problem of Swing oscillation

Oscillations of a swing pumped by a child is very familiar to us. Every time the swing passes through its lowest point the child pumps it over and again. The dynamics of weightless rod with a point mass sliding along the length mimics like a pendulum swing whose length varies periodically with time. The motion of the swing governed by the dynamical system written as [5]:

$$\frac{d}{dt}\left[ml^2\frac{d\theta}{dt}\right] + \gamma l^2\frac{d\theta}{dt} + mgl\sin\theta = 0 \tag{3}$$

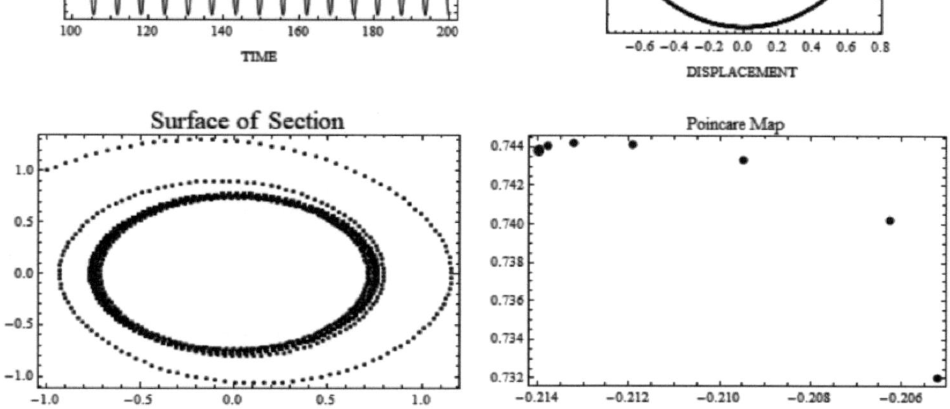

Figure 2.
Time-series plot, phase plot, surface of section and Poincaré map for periodic motion for F = 0.2.

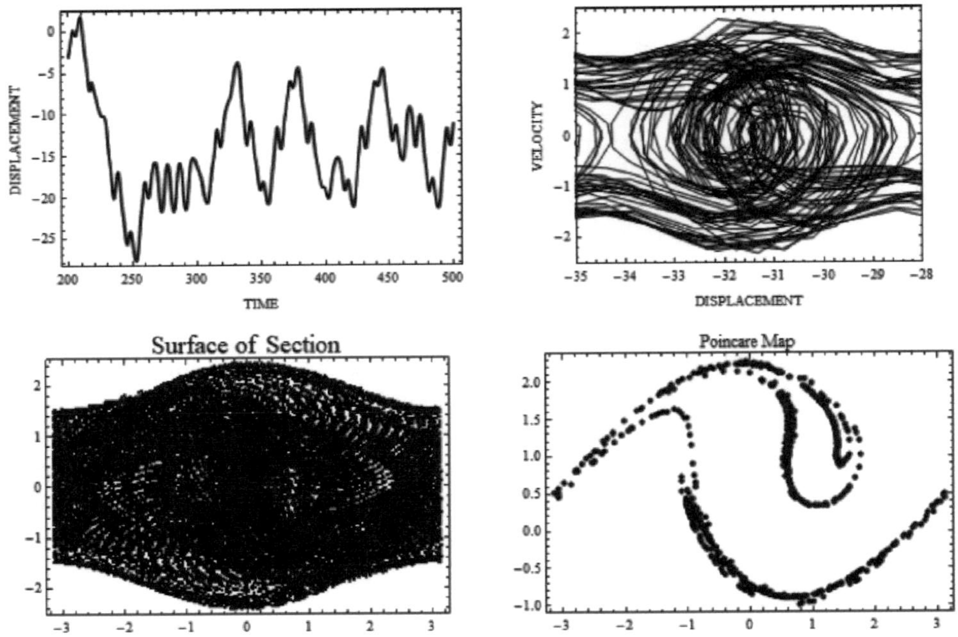

Figure 3.
Time-series plot, phase plot, surface of section and Poincaré map for periodic motion for $F = 0.8$.

where m is the mass, l is the length, θ is the angle made by the swing from the vertical position and g is the acceleration due to gravity. As the length of the swing varies periodically with time, one assumes

$$l = l_0 + a\phi(\Omega\, t) \tag{4}$$

where l_0 is the mean length of the swing and is constant, a and Ω are, respectively, the amplitude and frequency of excitation. The function $\phi(\Omega t)$ should be a periodic function of time. Then, by introducing the following dimensionless parameters and variables

$$\tau = \Omega\, t, \;\; \varepsilon = \frac{a}{l_0}, \;\; \Omega_0 = \sqrt{\frac{g}{l_0}}, \;\; \omega = \frac{\Omega_0}{\Omega}, \;\; \beta = \frac{\gamma}{m\Omega_0},$$

equation of motion of the swing in dimensionless form written as:

$$\ddot{\theta} + \left[\frac{2\varepsilon\dot{\phi}(\tau)}{1 + \varepsilon\phi(\tau)} + \beta\omega\right]\dot{\theta} + \frac{\omega^2}{1 + \varepsilon\phi(\tau)} \sin\theta = 0 \tag{5}$$

where $\{.\}$ in Eq. (5) corresponds to differentiation with respect τ.

Since $\phi(\tau)$ is a periodic function, we may take $\phi(\tau) = A \sin(\lambda\tau)$ and thus the foregoing equation may be rewritten as:

$$\ddot{\theta} + \left[\frac{2k\lambda \cos(\lambda\tau)}{1 + k\sin(\lambda\tau)} + \beta\omega\right]\dot{\theta} + \frac{\omega^2}{1 + k\sin(\lambda\tau)} \sin\theta = 0 \tag{6}$$

where $k = \varepsilon A$.

For stability of motion of the swing a linear stability analysis is applied. We may write Eq. (6) as the following two first order equations:

$$\begin{aligned} \dot{\theta} &= u \equiv f(\theta, u) \\ \dot{u} &= -\left[\frac{2k\lambda\cos(\lambda\tau)}{1+k\sin(\lambda\tau)} + \beta\omega\right]u - \frac{\omega^2}{1+k\sin(\lambda\tau)}\sin\theta \equiv g(\theta, u) \end{aligned} \tag{7}$$

The Jacobian for the above system may be written as,

$$J = \begin{pmatrix} 0 & 1 \\ -\dfrac{\omega^2}{1+k\sin(\lambda\tau)}\cos\theta & -\left[\dfrac{2k\lambda\cos(\lambda\tau)}{1+k\sin(\lambda\tau)} + \beta\omega\right] \end{pmatrix}$$

When an external periodic force $F\cos(\vartheta\tau)$ is applied to pump the swing, final form of the equation of motion stands as

$$\ddot{\theta} + \left[\frac{2k\lambda\cos(\lambda\tau)}{1+k\sin(\lambda\tau)} + \beta\omega\right]\dot{\theta} + \frac{\omega^2}{1+k\sin(\lambda\tau)}\sin\theta = F\cos(\vartheta\tau) \tag{8}$$

2.3 Regular and Chaotic motion of the swing

The swing, Eq. (8), oscillates in regular motion for significant contribution of friction, (i.e. when the frictional coefficient β has sufficiently higher value) and it is in chaotic motion in case of small friction and higher values of driving force. **Figures 4** and **5** showing the case of regular motion.

When the frictional contribution is insignificant, swing oscillations are chaotic and unpredictable. **Figure 6** stands for such chaotic motion of the swing when $\beta = 0$.

Figure 7 show chaotic oscillation when β is not zero but small. Surface of section and Poincare map shown in this figure are interesting showing typical chaotic behavior.

We may thus conclude that the swing oscillates smoothly when the frictions are higher but for no friction or insignificant friction, swing oscillations would be

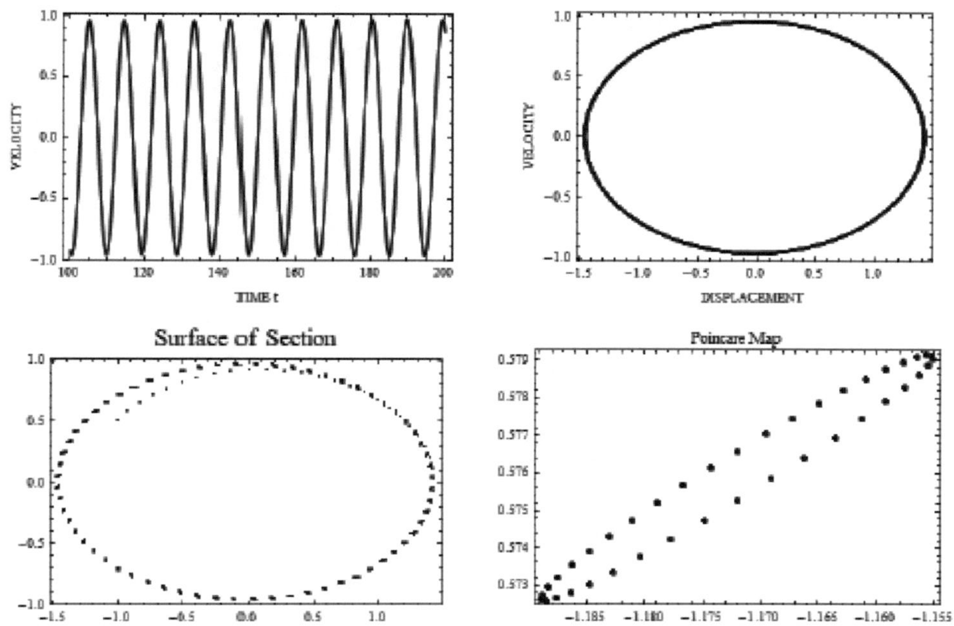

Figure 4.
A time-series and phase plots and plots of surface of section and Poincaré map for regular motion of the swing for $F = 0.8,\ \beta = 0.5,\ k = 0.1,\ \lambda = 0.05,\ \omega = 1,\ \vartheta = 2/3$.

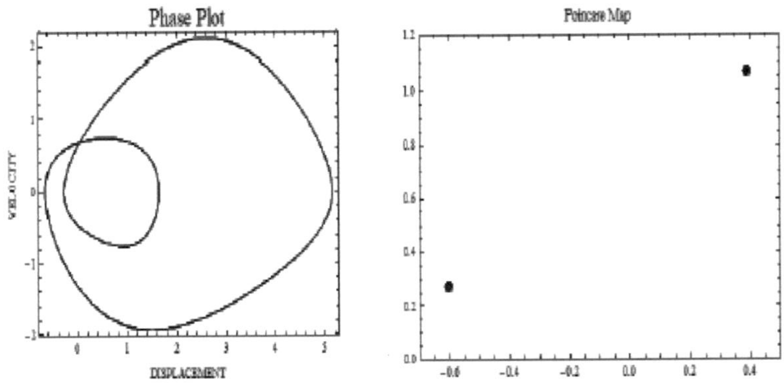

Figure 5.
Regular two periodic motion of the swing when $F = 0.9,\ \beta = 0.5,\ k = 0.3,\ \lambda = 1,\ \omega = 1,\ \vartheta = 2/3$.

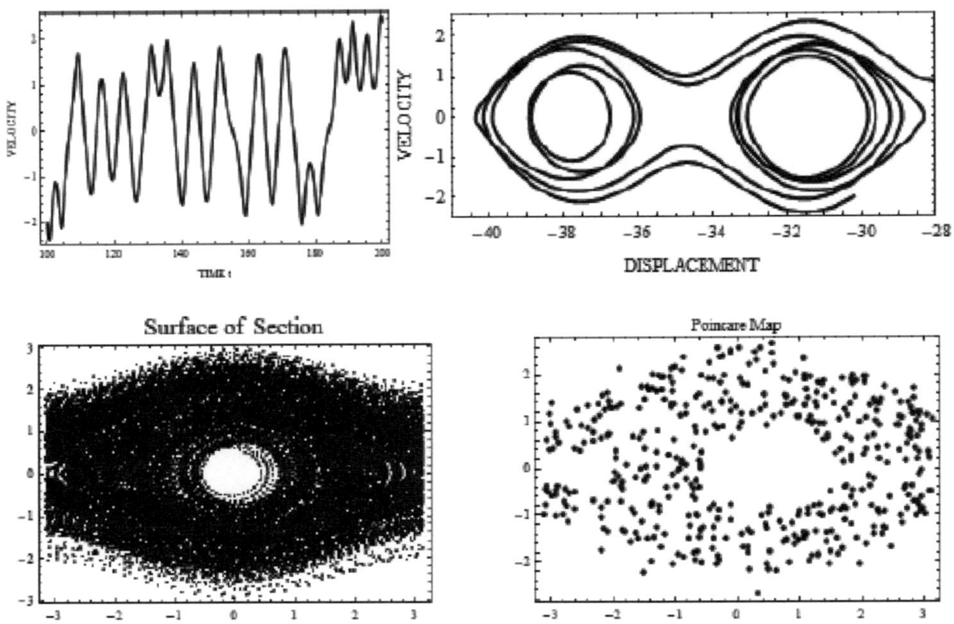

Figure 6.
Chaotic oscillation of the swing when $F = 0.2,\ \beta = 0,\ k = 0.1,\ \lambda = 0.05,\ \omega = 1,\ \vartheta = 2/3$.

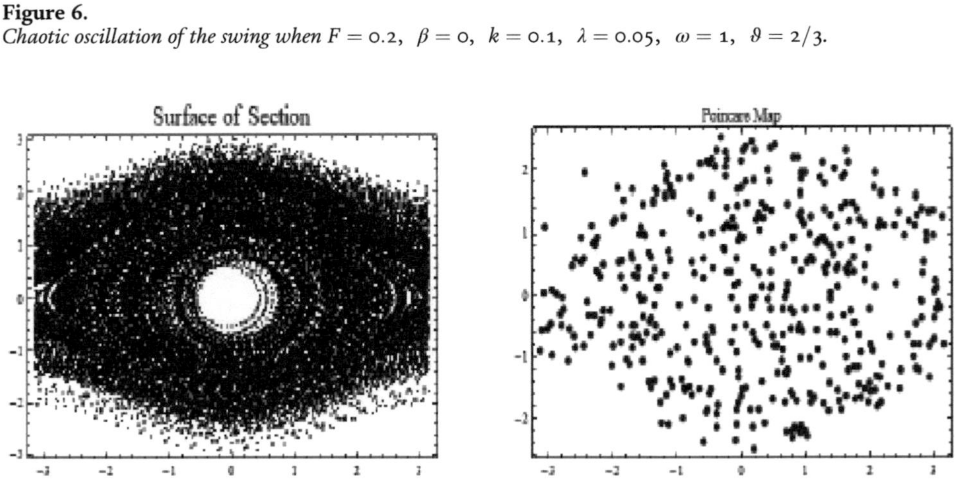

Figure 7.
Chaotic oscillation of the swing when $F = 0.2,\ \beta = 0.01,\ k = 0.1,\ \lambda = 0.05,\ \omega = 1,\ \vartheta = 2/3$.

chaotic or unpredictable. In such a case whirling, overturn or any unpredictable situation may happen.

Beside the application of bifurcation diagram, phase plot and Poincare surface of section technique, we introduce the idea of Lyapunov characteristic exponents (LCE), correlation dimension and topological entropy which provide further insight of a complex dynamical system. In the following section, we analyze the complexity of Prey-predator system using such tools.

3. Complexity in prey-predator system with Allee effect

In recent years many type of predator-prey problems, originated in Biological sciences, investigated which depend on various environmental and social conditions, [8–11]. Some problems solved by the application of Allee effect, which is an interesting phenomenon, to some predator-prey systems appear to be very interesting, [12–16]. The Allee effect on prey-predator system is a phenomenon in biology which characterizes certain correlation between population size or density and the mean individual fitness of a population or species. In the following study we investigate the complexity in a predator – prey problem with the Allee effect.

3.1 Discrete prey-predator model

A model for the prey-predator problem with Allee effect can represented as

$$\begin{aligned} X_{n+1} &= X_n + rX_n(1 - X_n)(1 - \exp(-\varepsilon X_n)) - aX_nY_n \\ Y_{n+1} &= Y_n + aY_n(X_n - Y_n)\left(\frac{Y_n}{\mu + Y_n}\right) \end{aligned} \tag{9}$$

where X_n and Y_n refers to the density of prey and predators. Further, r correspond to the growth rate parameter of the prey population and a the predation prameter. Here,

- $1 - \exp(-\varepsilon X_n)$ stands for mate finding Allee effect on prey population, here ε is defined as the Allee effect constant and the term
- $\frac{Y_n}{\mu + y_n}$ stands for the Allee effect on predator and here, μ is the Allee effect constant. Bigger μ means the stronger the Allee effect on predator population.

For assumed values of parameters $a = 2.0$, $r = 2.4$, fixed points of system (9) are obtained, approximately, as $P_1^*(0,0)$, $P_2^*(1,0)$, $P_3^*(0.545455, 0.545455)$ and by using stability analysis, we find all are unstable.

3.2 Bifurcation diagrams

The phenomena of bifurcation provide a qualitative change in the behavior of a system during evolution. Such a change occurs when a particular parameter is varied while keeping other parameters constant. Bifurcation diagram shows the splitting of stable solutions within a certain range of values of the parameter. During the processes of bifurcation, one observes different cycles of evolution which leading to the chaotic situation. Phenomena like bistability, periodic windows within chaos etc. may also be observed for some systems. A bifurcation can be taken as a tool to analyze the regular, chaotic as well as complexity within the system. For $a = 2.0$ and $1.8 \leq r \leq 2.4$, **Figure 8** shows bifurcation of system (9), where some interesting phenomena observed that the system is not producing a period doubling bifurcation scenario which is very common for many nonlinear systems.

3.3 Numerical simulations

Simulation of the foregoing models provides,

3.3.1 Attractors

Keeping parameters a and r fixed viz., $a = 2.0, \ \ r = 2.4$, attractors for different cases are obtained through numerical technique [16], and shown in **Figure 9**.

Looking plots of attractor of **Figure 9**, one finds a chaotic attractor, figure (a), when Allee effect is not in consideration, for $a = 2.0, \ \ r = 2.4$. But, the application of Allee effect to either of the population or to both population, system returned to regularity, e.g. figures (b), (c) and (d) are no more chaotic. This also follow from the plots of LCEs given below.

3.3.2 Lyapunov exponents (LCEs)

The phase space dynamics of a nonlinear chaotic physical system is very complex in general. One of the important feature of such a system is its sensitivity to initial conditions i.e., two very nearby trajectory in phase space show divergence exponentially. Such divergence are characterized by LCEs. To indicate chaotic and regular evolution, an appropriate measure is to find Lyapunov exponents (LCEs) which are obtained for different cases by using appropriate procedure. Plots of LCEs are shown in **Figure 10**.

3.3.3 Correlation dimension

Lorenz attractor provides an example of a fractal object with noninteger dimension. The correlation dimension permits us to quantify the space filling property and provides the measure of dimensionality of the chaotic attractor. It is expressed as

$$D^c = \frac{d \log C(R)}{d \log (R)}$$

where $C(R)$ is defined as

$$C(R) = \frac{1}{n(n-1)} \sum_{i=1}^{n} \sum_{j=1}^{n} [\Theta(R - \|x[i] - x[j]\|)]$$

corresponds to the correlation sum and is a measure of total number of points contained within a hypersphere of radius R as a function of R normalized to the total number of points squared. Using the algorithm [17, 18], the correlation dimension can be determined from the scaling region found in the plot of $\log C(R)$ as a function of $\log (R)$. For the Lorenz system with parameters $\sigma = 10, \ \ \rho = 28$ and $b = 8/3$, the correlation dimension D^c is found to be 2.069. The correlation dimension of the chaotic attractor **Figure 9(a)** is found to be $D^c = 0.571$ (**Figure 11**).

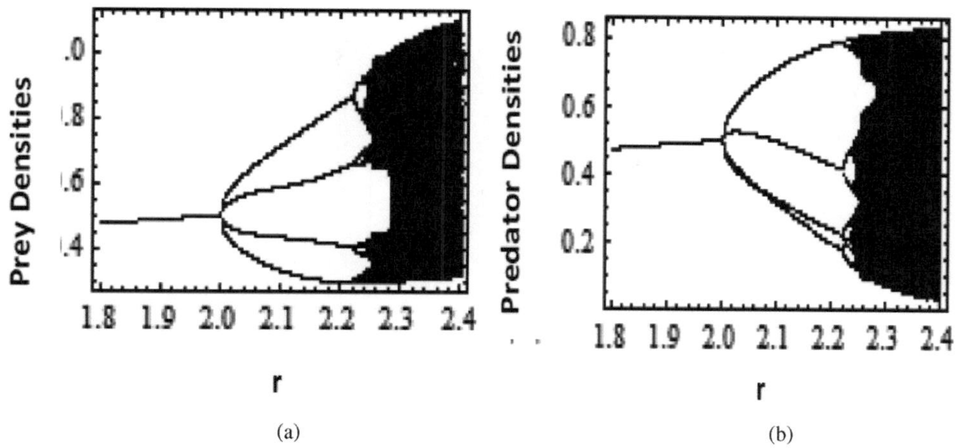

Figure 8.
Bifurcation diagram of system (9), (a) Prey densities, (b) Predator densities for $a = 2.0$ *and* $1.8 \leq r \leq 2.4$.

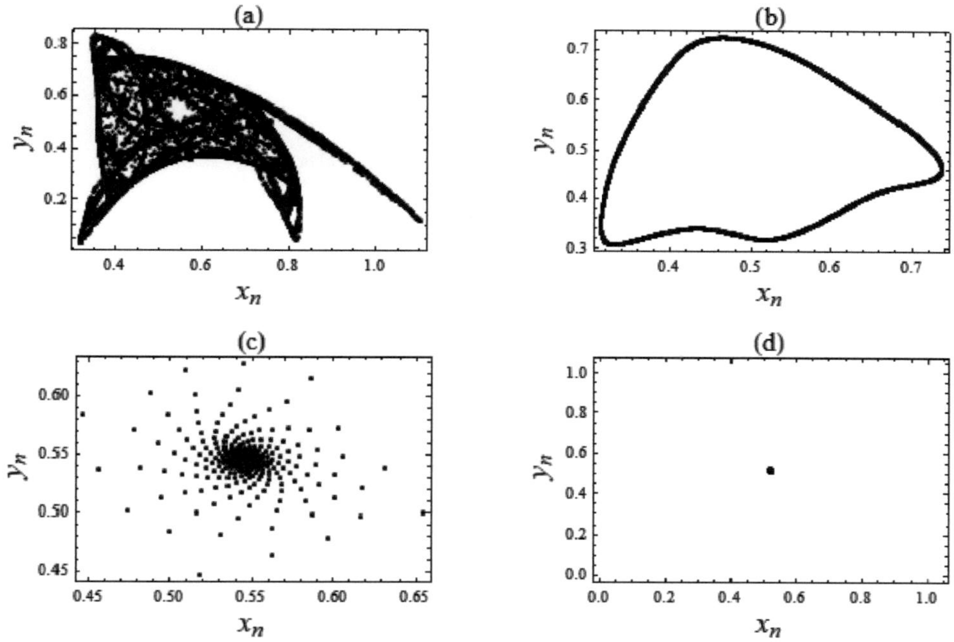

Figure 9.
Plots of regular and chaotic attractors for $a = 2.0$ *and* $r = 2.4$*; (i) plot (a) without Allee effect, (ii) plot (b) with Allee effect on prey only,* $\varepsilon = 4.5$ *, (iii) plot (c) Allee effect on predator only mu* $= 0.1$*, and (iv) plot (d) Allee effect on prey as well as on predator,* $\varepsilon = 4.5$*,* $\mu = 0.1$.

3.3.4 Topological entropies

As explained in the beginning, topological entropy measures the complexity of the system. More topological entropy implies system is more complex. Presence of complexity does not mean the system is chaotic and vice versa. In **Figure 12**, we have plots of topological entropy for different cases. In figure (a), topological entropy increases for $r > 2$ but bifurcation diagrams and calculations of LCEs indicate the system is regular within $2.0 \leq r \leq 2.2$. Similar observation can be made looking at figures (b) and (c). In figure (d) one finds no fluctuations of topological entropy, it establishes a steady state situation.

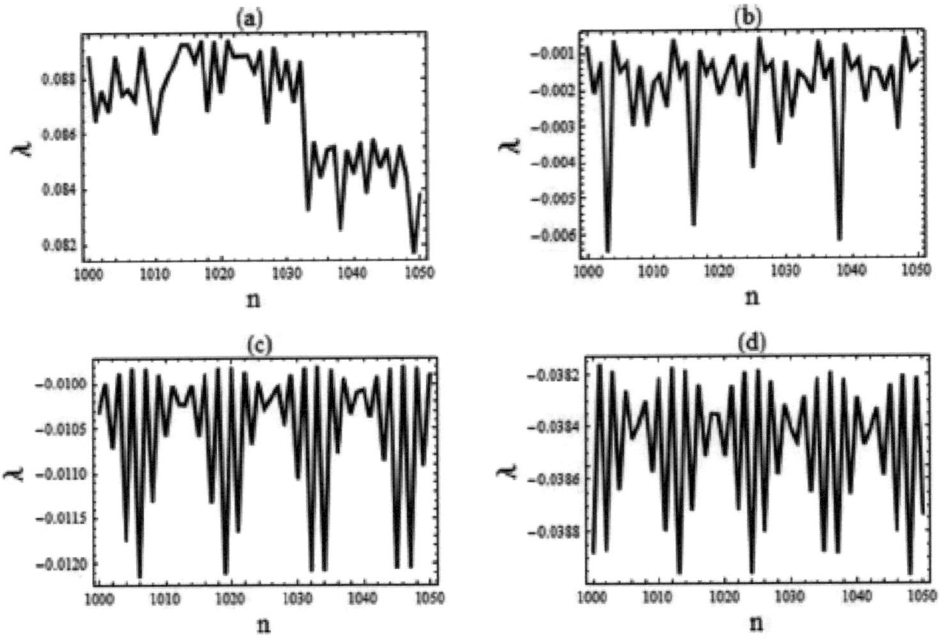

Figure 10.
Plots of Lyapunov exponents for $a = 2.0$, $r = 2.4$ *and (i) figure (a) without Allee effect, (ii) figure (b) with* $\varepsilon = 4.5$, $\mu = 0$, *(iii) figure (c) Allee effect on predator only with* $\mu = 0.1$, *(iv) Allee effect on both populations* $\varepsilon = 4.5$, $\mu = 0.1$.

The results obtained through bifurcation plots, **Figure 8**, and those of LCEs plots, **Figure 10**, show that the Allee effect stabilize the motion from chaos to regularity. The correlation dimension of the chaotic attractor is obtained as $D^c \cong 0.571$. Through this study we find the existence of complexity within the system, even when system behavior is regular, we find significant amount of increase in topological entropy. This implies the fact that the system may be regular but may exhibit complexity.

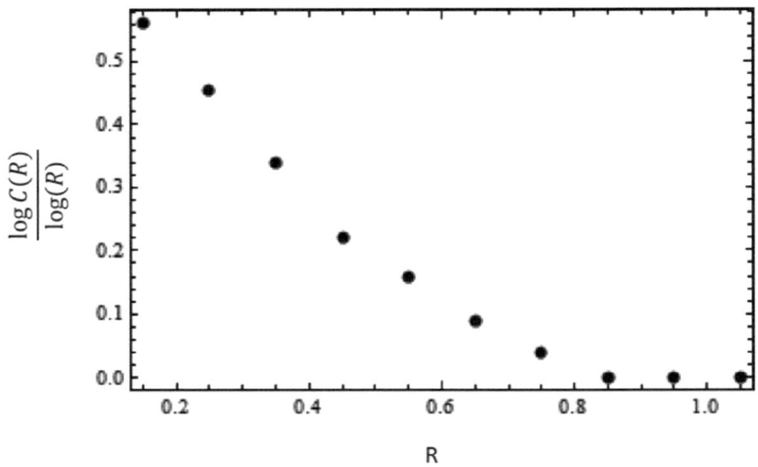

Figure 11.
Plot of correlation integral data.

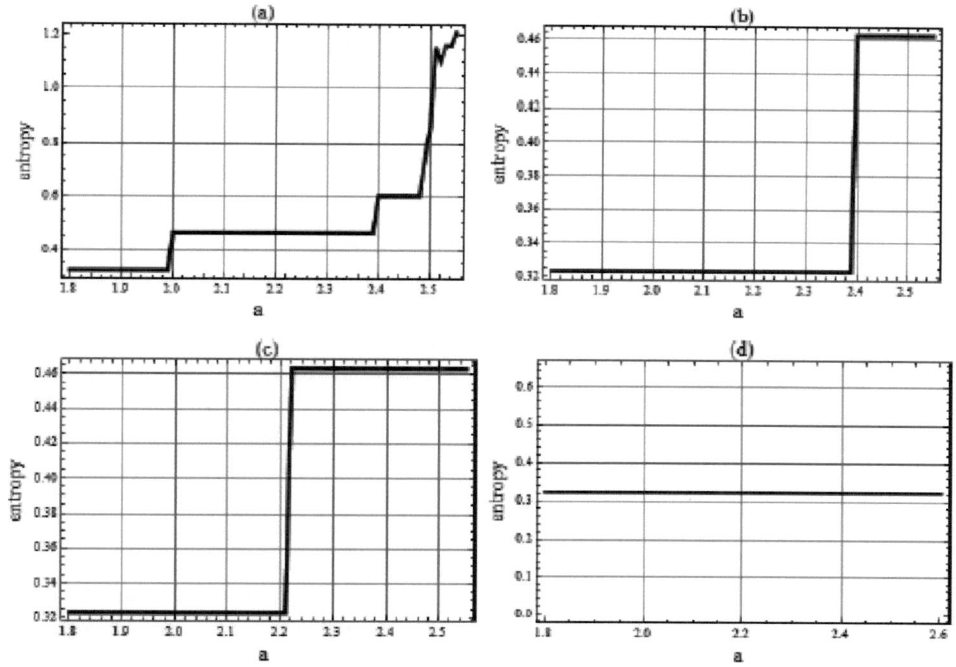

Figure 12.
Plots of topological entropies for $a = 2.0$ *and* $1.8 \leq r \leq 2.6$ *: (i) figure (a) with no Allee effect, (ii) figure (b) when* $\varepsilon = 4.5, \ \ \mu = 0$*, figure (c) with Allee effect on predator only* $\mu = 0.1$ *, (iv) when* $\varepsilon = 4.5, \ \ \mu = 01$.

4. Recurrence plot

Natural system exhibits periodicities and also irregular cyclicities. Usually measures such as Lyapunov characteristic exponent (LCE), correlation dimension, Kolmogorov- Sinai (KS) entropy etc., have been used to characterize the complexity of observed nonlinear dynamical behavior of a system. But the analysis based on application of the foregoing tools inherently assumes the system to be noise free and stationary. An alternative framework based on the idea of recurrence plot was introduced in [19] for visualization of the dynamical behavior of a system in phase space and subsequently the formalism has been extended to quantify the recurrence plots to unravel the observed complexities i.e., regular, quasi-periodic, chaotic transition etc. For a discrete time series P with N data points such that

$$P : \{x_1, x_2, x_3, \cdots, x_N\}, \tag{10}$$

where $x_i, i = 1, 2, \ldots \ldots, N$ refers to observed values at time $t_1, t_1 + \Delta t, \ldots \ldots, t_1 + n\Delta t$. If the system has true dimension m , a sequence of vectors may be constructed from the time series as:

$$\mathbf{X_i} = \left(x_i, x_{i+\tau}, x_{i+2\tau}, \cdots, x_{i+(m-1)\tau}\right); \ \ i = 1, 2, \cdots, n; \ \ n = N - (m-1)\tau \tag{11}$$

where τ corresponds to time lag or delay and m - the embedding dimension of the phase space. By considering the distances in m- dimensional reconstructed points, we construct a recurrence plot (RP). In fact RP is an $n \times n$ symmetrical array where a *dot* is marked at a point (i,j) if $\mathbf{X_i}$ is close to another point $\mathbf{X_j}$.
We may write

$$R_{i,j}(r) = \Theta(r - \|\mathbf{X_i} - \mathbf{X_j}\|); \quad i,j = 1, 2, \cdots, n \tag{12}$$

where Θ is a Heavyside function, r is a small threshold distance between neighboring points and $\|.\|$ is an Euclidean norm.

A characteristic pattern emerges in RPs which characterizes a dynamical system. The method of RP is suitable for both stationary and non-stationary dynamical system. Since a trajectory may return to a point or close to it in phase space, the deterministic dynamical system shows recurrent behavior and RP therefore exhibits both horizontal and vertical lines. For a stochastic dynamical system, such lines in RP are of very small size and in fact appear by chance. Therefore the distribution of such points appear to be homogeneous. In case of periodic system the RP is filled with longer diagonal lines. Various measures that quantify RPs are mainly, *RR, DET, ENT, DIV, LAM, TT* which refers to density of recurrence points, determinism, divergence, entropy, laminarity and trapping time respectively. For a periodic system diagonal lines are longer which for chaotic system RP shows broken short lines. Recently [20] has provided a very useful description of applying RPs and recurrence quantification analysis to unravel the complex dynamics of general problem of three species interaction in ecology. In the following section, we extend the analysis of complexity to the problem associated with neuronal dynamics, an area of current interest in neuro-bio-science [21]. We however restrict ourselves to only RPs to supplement the analysis of complexity using phase portrait, bifurcation diagram etc.

5. Regular and Chaotic neuronal dynamics

The neuronal communication is known to be mediated by electrical pulses called spikes. Studies of various spiking patterns reveal nonlinear characteristics of slow-fast neuronal dynamics. A considerable amount of information regarding neuronal activity has been obtained by studying the dynamics of spiking pattern [22]. The phenomenon of tonic firing, mixed mode (bursting and spiking) etc., are typical responses exhibited by an excitable neuron [23]. Cortical neurons have been reported to show tonic bursting wherein the neurons periodically switches between firing state and resting state. The mixed mode firing is observed in mammalian neocortex [24]. Spike generation in fact depends on the firing threshold and the stimulus intensity. In recent years, the perception regarding constancy of neuron's firing threshold has changed to dynamic [21]. In this work, we first briefly introduce the Fitzhugh-Nagumo model (FHN) that have been proposed for spike generation like well known Hodgkin-Huxley model. It is however to be noted that FHN model reproduce the experimental results less accurately. Our interest in FHN model emanates mainly due to its showing complex spiking pattern even though it is mathematically simple. The basic FHN model assumes the threshold to be constant. We also study the changes caused in the spiking pattern as a result of time varying threshold. This study assumes significance as such a model may throw insight into the model the dynamics of cortisol secretion from hypothalamus [25]. It is to be noted that the neuronal firings may take place at regular interval or randomly due to inherent mechanism or may be due to its interaction with the neighborhood neurons or result of exogenous stimulus [26–28].

5.1 Basic dynamics of FHN model

The FHN model describes the interaction between the voltage v across the axon membrane driven by input current I and the recovery variable w. The recovery variable w is the result of mainly the reflecting outward potassium current (K^+) that results in hyperpolarization of the axon after each spike occurrence. We may write the FHN model equation as [28–30]:

$$\begin{aligned} \frac{dv}{dt} &= \alpha v(\beta - v)(v - b_0) - \sigma w + I \\ \frac{dw}{dt} &= \varepsilon(v - \delta w) \end{aligned} \tag{13}$$

where $\delta > 0$ and the parameter $\alpha > 0$ scales the amplitude of the membrane potential v, and ε is used here to control the recovery variable w with respect to action potential v. The parameter b_0 i.e., $(0 < b_0 < 1)$, corresponds to the threshold value that controls the excitable behavior of the neuron. Also β and σ are constants for the system.

In our analysis of Eq. (13), we take $\beta = \sigma = 1$ and for the case of no external input current, $I = 0$ the dynamical system (13) has three equilibrium points or fixed points (v_e, w_e) as:

$$E_1 = (0, 0),$$

$$E_{2,3} = \frac{(1 + b_0)}{2} \pm \frac{\sqrt{(1 - b_0)^2 - \frac{4}{\alpha\delta}}}{2}.$$

It may be noted that if

$$(1 - b_0)^2 - \frac{4}{\alpha\delta} < 0, \tag{14}$$

then the system possess E_1 as the only equilibrium point.

Further defining

$$h(v) = v(1 - v)(v - b_0) \tag{15}$$

the Jacobian matrix J of the system may be written as:

$$J = \begin{vmatrix} \alpha h'(v) & -1 \\ \varepsilon & -\varepsilon\delta \end{vmatrix}. \tag{16}$$

For the equilibrium or fixed points (v_e, w_e), the eigenvalues $\lambda_{1,2}$ of the Jacobian matrix are given by

$$\lambda_{1,2} = \frac{-(\varepsilon\delta - \alpha b_1) \pm \sqrt{(\varepsilon\delta - \alpha b_1)^2 - 4\varepsilon(1 - \alpha b_1 \delta)}}{2} \tag{17}$$

where $b_1 = h'(v_e)$.

Therefore (1) if $\alpha b_1 \delta < 1$, the equilibrium point (v_e, w_e) is asymptotically stable if $\alpha b_1 < \varepsilon\delta$, a repellor if $\alpha b_1 > \varepsilon$, (2) if $\alpha b_1 \delta > 1$ the equilibrium point is a saddle point and (3) if $\alpha b_1 \delta = 1$ then the equilibrium point is stable (unstable) if $\alpha b_1 < \delta\varepsilon (\alpha b_1 > \delta\varepsilon)$.

In case the parameters of the system are such that condition Eq. (14) holds then using Eq. (17) we find the equilibrium point E_1 i.e., origin, to be asymptotically stable if

$$Re\left[-(\alpha b_0+\varepsilon\delta)\pm\sqrt{(\alpha b_0+\varepsilon\delta)^2-4\varepsilon(\alpha\delta b_0+1)}\right]<0. \quad (18)$$

Based on Routh's criteria we may write equivalently

$$\alpha b_0+\varepsilon\delta>\quad 0,\qquad \alpha\delta b_0+1>0. \quad (19)$$

Therefore if origin is the only fixed point of the system, then following [30], it is observed that the system has no limit cycle if $\left(b_0-\frac{1}{2}\right)^2+\frac{3}{4}-\frac{3\delta\varepsilon}{\alpha}<0$. This result however assumes $\alpha b_0+\varepsilon\delta>0$. In case $\alpha b_0+\varepsilon\delta<0$, the origin becomes unstable and the system can be shown to exhibit one stable limit cycle. It is noted here that on varying the threshold parameter b_0, the system may exhibit Andronov-Hopf bifurcation when $\alpha b_0+\varepsilon\delta=0$ as per Eq. (18) and at this point the origin of the system becomes unstable causing a bifurcation to at least one stable limit cycle. It is to be noted that gaps exist in parameter space when origin is the only asymptotically fixed point and where limit cycles may exists. Interestingly, we numerically show the existence of bistable behavior [18, 30] in terms of occurrence of double cycle bifurcation by taking $\varepsilon=0.015,\ \delta=3.5,\ \alpha=1.0$ and allowing the threshold value $b_0<0$ i.e., -0.044(**Figure 13**).

For the case of $I\neq 0$, the equilibrium points may be one, two or three and their stability may be analyzed following the foregoing analysis. Taking the parameter values: $a=0.06,\ b_0=0.50,\ \varepsilon=14$, as in [29], the phase portrait were obtained using numerical integration of the system, Eq. (13), for different I values (**Figure 14**). The appearance of limit cycle behavior is due to supercritical Hopf bifurcation and as a consequence of loss of stability of the unique equilibrium point that exist for $I<4.2$ [29]. **Figure 14** also suggest that the amplitude of limit cycles first increases and subsequently decreases with increase in values of I. At around $I\sim 12.45$ the second bifurcation occurs and system is led to a stable equilibrium. [29] has provided a detailed discussion on the richness of various bifurcation event as I is varied.

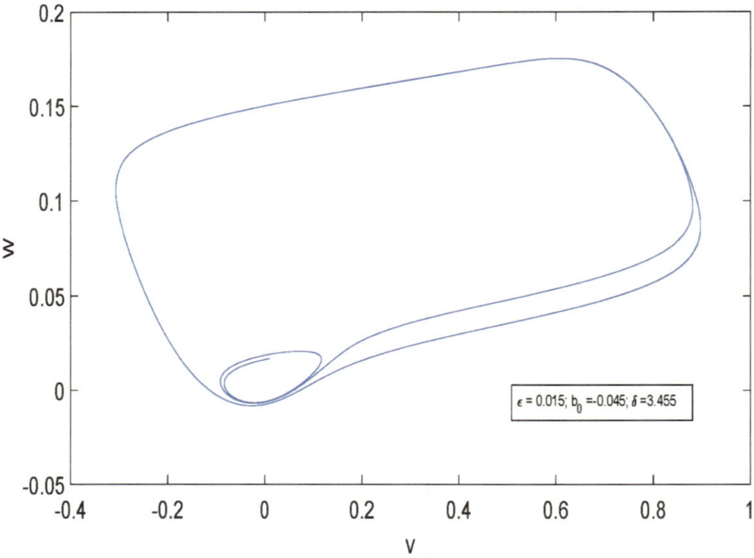

Figure 13.
Phase portrait of FHN system showing bistability between limit cycle and stable fixed point.

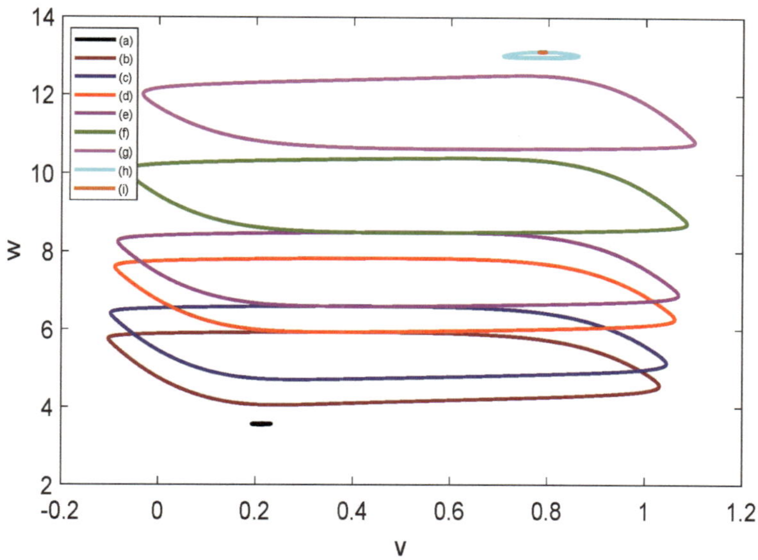

Figure 14.
Phase portrait of FHN system showing limit cycle for $4.2 \leq I \leq 12.45$. *(a)* $I = 4.23$, *(b)* $I = 4.8$, *(c)* $I = 5.5$, *(d)* $I = 6.8$, *(e)* $I = 7.5$, *(f)* $I = 9.5$, *(g)* $I = 11.75$, *(h)* $I = 12.42$, *and (i)* $I = 12.45$.

5.2 FHN neuron in the presence of external periodic electrical stimulation

Chaotic systems exhibit complexity and are sensitively dependent on initial condition of the system under investigation and also unpredictable. Chaos as a nonlinear phenomenon has attracted researchers from different disciplines e.g., physics, biology, ecology, neurobioscience etc. In this section, we investigate the effect of periodic electrical stimulation on the dynamics of an FHN system, Eq. (13).

The basic equation that governs the dynamics of FHN system in the presence of external periodic stimulation, $I(t)$, may be written as:

$$\frac{dv}{dt} = \alpha v(\beta - v)(v - b_0) - \sigma w + I(t),$$
$$\frac{dw}{dt} = \varepsilon(v - \delta w). \tag{20}$$

Here, we take the external periodic stimulation as given by $I(t) = \left[\frac{I_0}{2\pi\nu}\right] \cos(2\pi\nu t)$, where I_0, ν refers to the amplitude and frequency of the input stimulus. Further, we present the simulation results of the system, Eq. (20), by taking $\alpha = 10$, $\beta = 1$, $b_0 = 0.10$, $\delta = 0.25$, $\varepsilon = 1$, and $\sigma = 1$ and varying both I_0 and ν. The variation of both amplitude and frequency of the external periodic stimulus is found to result in the membrane potential v exhibiting regular or chaotic temporal behavior. The regular or periodic neuron spiking could be classified as $p : q$ phase-locking, where p and q corresponds to the number of spikes and number of periodic stimuli per unit response period. For instance **Figures 15** and **16** illustrates respectively the response of the neuronal spiking corresponding to 1 : 1 and 1 : 2 phase locked rhythm.

The response of the single FHN neuron to external periodic response could also be chaotic for certain values of the amplitude I_0 and frequency ν of driving stimulus i.e., i.e., $I_0 = 0.183$, $\nu = 0.1931$, as shown in **Figure 17**.

The observed dynamical transition from regular/periodic to chaotic of membrane potential v with increase in amplitude and frequency of external could be

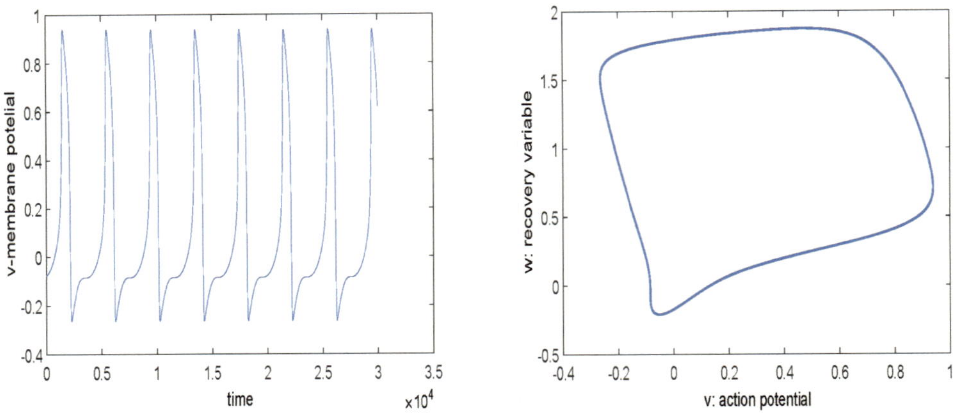

(a) (b)

Figure 15.
1 : 1 phase locking rhythm of spiking neuron. (a) Time series of membrane potential with $I_o = 0.1$, $\nu = 0.05$. (b) $v - w$ phase portrait with same parameters as in (a).

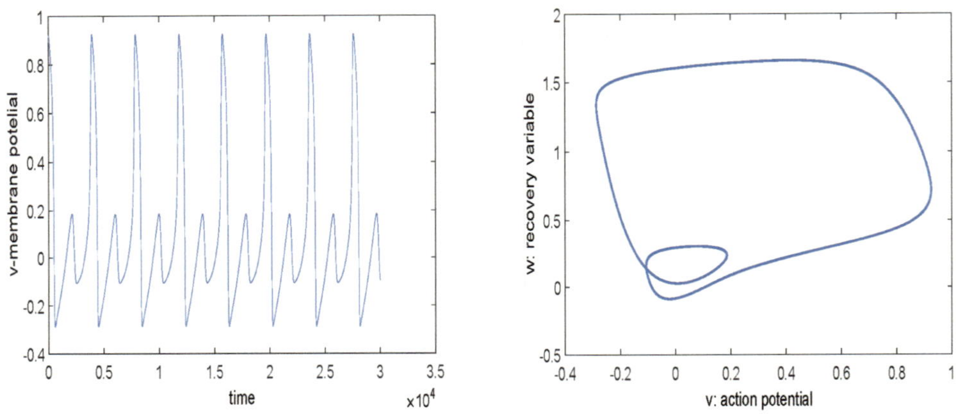

(a) (b)

Figure 16.
1 : 1 phase locking rhythm of spiking neuron. (a) Time series of membrane potential with $I_o = 0.1$, $\nu = 0.1015$. (b) $v - w$ phase portrait with same parameters as in (a).

further seen by constructing the RPs. The construction of RP, discussed earlier in section 4, involves the reconstruction of phase space using a time series of a dynamical variable, say the membrane potential v, based on the information regarding the delay parameter τ and the embedding dimension m. The delay parameter τ for the time series of v could be obtained using the method of mutual information (MI) [31] and the embedding dimension m may be determined using the algorithm of [32, 33]. The time series of **Figures 15–17** for the membrane potential v have been used to construct the RP shown in **Figure 18**. The change in spiking patterns caused by external periodic stimulation from regular to chaotic is well indicated in RP of almost equally spaced diagonal lines to irregularly occurring broken diagonal lines of varying length.

5.3 FHN neuron with time varying threshold

The dynamics of cortisol secretion from hypothalamus could be modeled using FHN system with time varying threshold [25]. Complexities of spike dynamics of FHN neuron has been earlier investigated in [34] incorporating the time varying

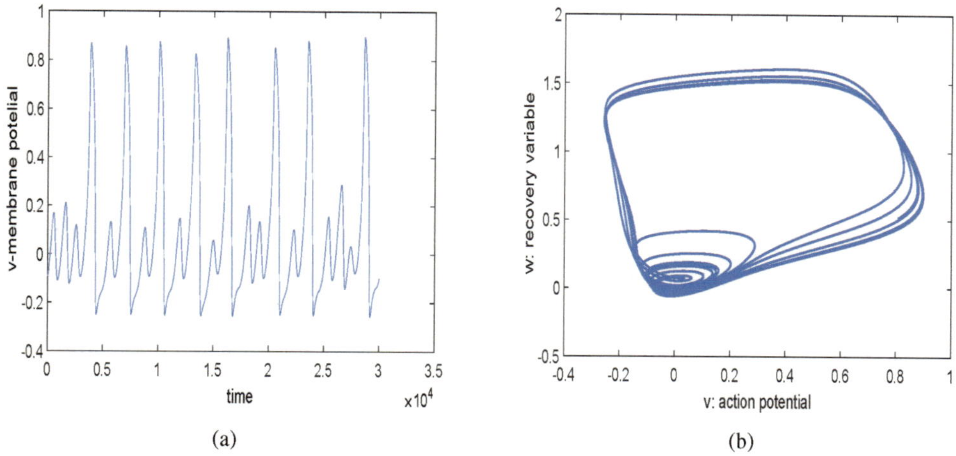

Figure 17.
1 : 1 phase locking rhythm of spiking neuron. (a) Time series of membrane potential with $I_o = 0.183$, $\nu = 0.1931$. *(b)* $v - w$ *phase portrait with same parameters as in (a).*

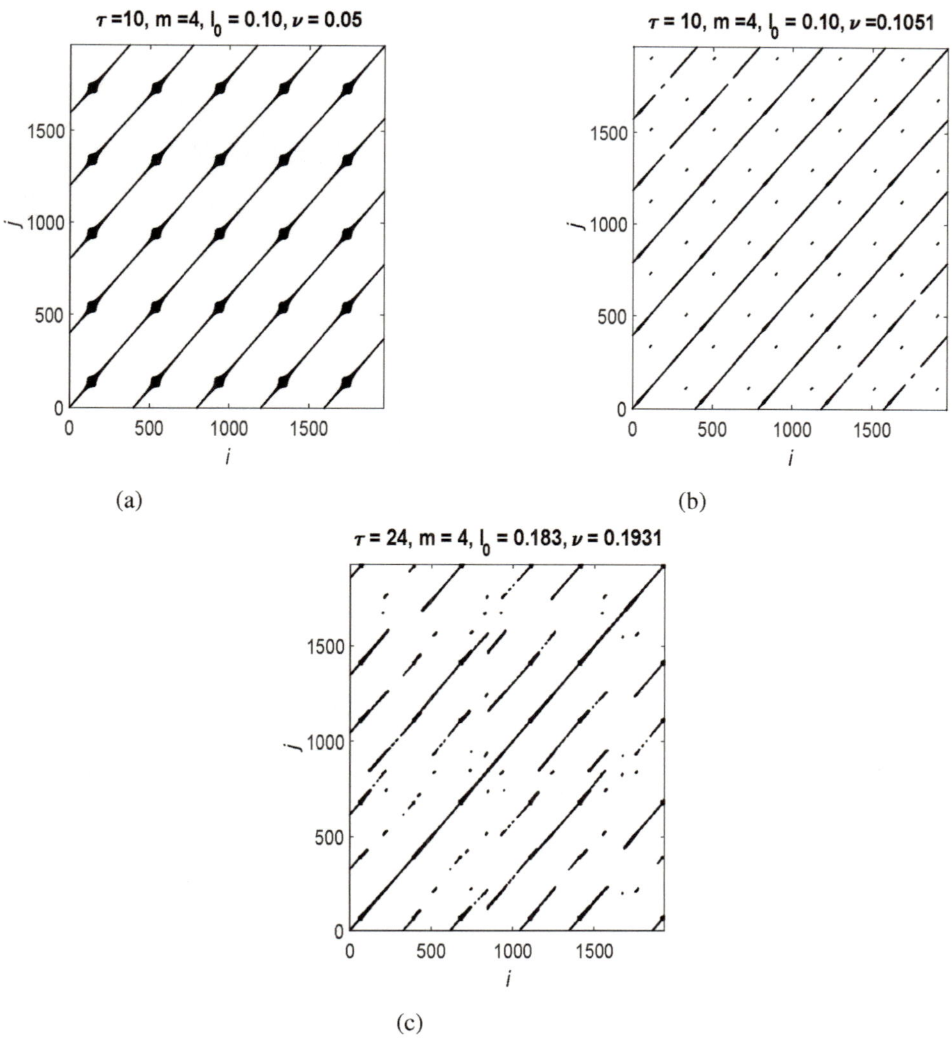

Figure 18.
RP of of the membrane potential with (a) 1 : 1 phase locking rhythm, (b) 1 : 2 phase locking and (c) chaotic rhythm.

threshold. Models of observed tonic firing and spike bursting were simulated by considering both periodic and noisy form of the threshold variation. The effect of a mixed mode threshold on the spiking FHN system was also investigated. Here we present results of bifurcation analysis of different states of neuronal firing of FHN neuron by considering a discrete form of the system [34, 35]. Following [34], the discrete form of the FHN system may be written as:

$$v_{n+1} = v_n + \Delta\alpha[-v_n(v_n - 1)(v_n - b_n) - w + I] \tag{21}$$

$$w_{n+1} = w_n + \Delta(v_n - \delta w_n) \tag{22}$$

where Δ refers to the integral step size and is treated here as a bifurcation parameter.

In case of mixed mode threshold variation, the membrane potential v exhibits a complex behavior as shown in bifurcation diagram (**Figure 19a**). It is readily observed that the temporal behavior is chaotic in the region $0.42 \le \Delta \le 0.68$. Thereafter windows of regular and chaotic regimes are observe till $\Delta = 0.8$ for $I = 0$. A slight increase in I changes the dynamics to a more complex behavior as shown in

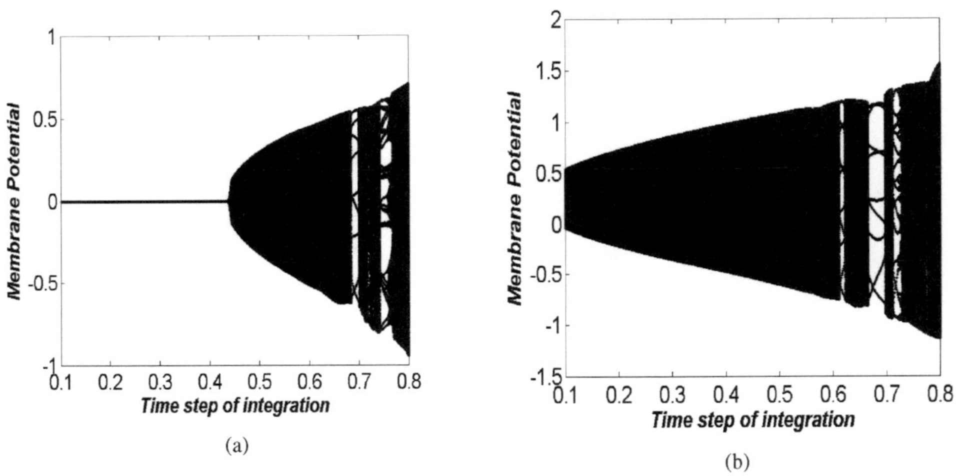

Figure 19.
Bifurcation of membrane potential. (a) I = 0; (b) I = 1.

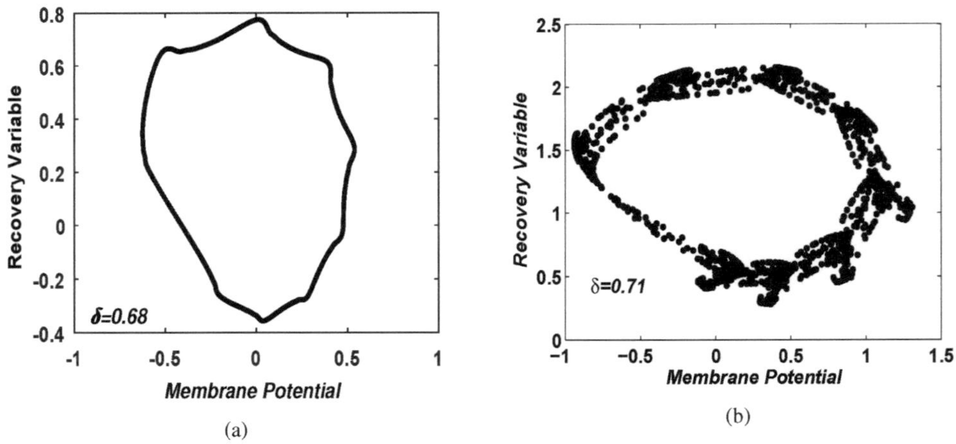

Figure 20.
Phase portrait showing quasiperiodic behavior for (a) I = 0 and chaotic behavior for (b) I = 1.

Figure 19b wherein windows of quasi-periodic behavior that sets sets in say at $\delta \sim 0.68$ for $I = 0$ (**Figure 20a**) make a transition to chaos as shown in **Figure 20b**.

In the following section, we study the complex dynamics in economics, an area actively pursued by researchers, by introducing another measure called 'multi-scale permutation entropy' by considering a nonlinear financial model.

6. Chaotic dynamics in finance model

In mainstream economics, economic dynamics has assume great importance in recent years in view of the availability of market and other data. Economic dynamics has therefore influenced both micro- and macroeconomics. Therefore lot of research output has poured in explaining irregular micro-economic fluctuation, erratic business cycles, irregular growth and aperiodic behavior of economic data etc. Nonlinear systems provides an alternative simple and deterministic framework that easily can explain aperiodic or chaotic behaviors of various financial systems. One of the important features of nonlinear system is that the irregular/chaotic behavior supports an endogenous mechanism for the observed complexity in economic time series. As a result nonlinear dynamic framework has been applied to economic modeling and several examples are available in [36–46].

In the present work we revisit the synthetic chaotic financial model discussed in [45, 46] which is based on interest rate, investment demand and price index as dynamical variables. We numerically explored and analyze the complexity of the model using the multiscale entropy (MPE) frame work. In this section, we briefly describe the chaotic financial model and its basic characteristics. We also outlines the procedure of MPE for analyzing the complexity of the finance model.

6.1 Chaotic financial model

We consider a dynamic finance model composed of three coupled first order differential equation. This model describes the temporal evolution of the state variables viz. the interest rate X, the investment demand Y and the price index Z. The model is described as [39]:

$$\begin{aligned} \frac{dX}{dt} &= Z + (Y - a)X, \\ \frac{dY}{dt} &= 1 - bY - X^2, \\ \frac{dZ}{dt} &= -X - cZ. \end{aligned} \tag{23}$$

Here a, b and c are positive constants and represent the saving amount, cost per investment and elasticity of demand of the commercial markets. First equation appears, representing the changes in X, as a result of contradiction in the investment market and structural adjustment from the goods prices. Second equation representing the changes in Y appears due to proportionality to the rate of investment and also to an inversion of the cost of investment and interest rate. The third equation emerges due to contradiction between supply and demand in commercial markets which is influenced by interest rates.

Eq. (23) has been numerically integrated using fourth-order Runge-Kutta method to obtain the time series of the dynamic variables X, Y and Z, shown in **Figure 21** with $a = 3.0$, $b = 0.1$, $c = 1.0$ and initial condition $(X_0, Y_0, Z_0) = (2, 3, 2)$. Similar choice of the parameter were made in [45].

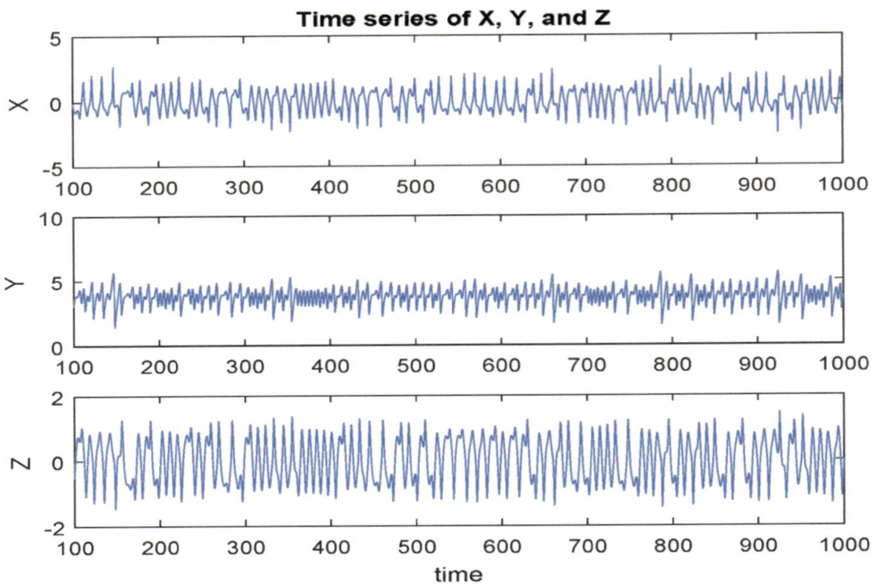

Figure 21.
Temporal evolution of Finance model.

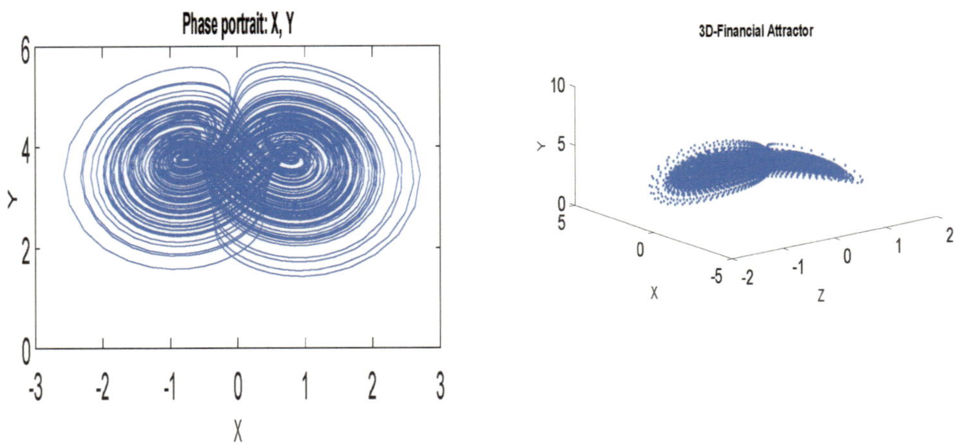

(a)

(b)

Figure 22.
(a) Phase portrait of Finance model, (b) 3-D attractor of the Finance model.

The representation of two dimensional phase portrait $(X, \ Y)$ and the attractor are shown in **Figure 22**. Obviously they represent chaotic dynamics of the temporal behavior shown in **Figure 22**.

6.2 Complexity analysis using multiscale permutation entropy (*MPE*) method

The multiscale permutation method involves two steps. A "coarse graining" is applied first to a time series $X_i, \ i = 1, N$ to construct a consecutive coarse-grained time series. The coarse-grained process involves averaging a successively increasing number of data points in non-over lapping windows. The elements of each of the coarse grained time series $y_j^{(s)}$ is computed as,

$$y_j^{(s)} = \frac{1}{s} \sum_{i=(\ j-1)\tau+1}^{js} X_i \tag{24}$$

where $1 \leq i \leq N/s$ and s defines the scale factor. Each time series length is of size that is an integral multiple of N/s. For $s = 1$, the coarse-grained time series is just the original time series.

The second step involves the computation of permutation entropy [47] for each of the coarse-grained time series. For a coarse-grained time series y_j we first consider the series of vector of length m, and obtain $S^m(n) = [y_n, y_{n+1}, \cdots, y_{n+m-1}]$, $1 \leq n \leq (N/s) - m + 1$. Subsequently, $S^m(n)$ is arranged in an increasing order viz., $\left[y_{n+j_1+1} \leq y_{n+j_2+1} \leq \cdots \leq y_{n+j_n+1}\right]$. For m different numbers, there will be $m!$ possible order patterns/structures Π which are termed as permutations. If $f(\Pi)$ denotes the frequency of order pattern Π, then the relative frequency and hence the probability $p = f(\Pi)/(N/s - m + 1)$. The permutation entropy $H(m)$ therefore is given by

$$H(m) = -\sum_{\Pi=1}^{m!} p\,(\Pi)\ ln\ (\Pi). \tag{25}$$

The maximum value of $H(m)$ is $log\ (m!)$ thus showing all permutations to have equal probability. Also, the time series is termed as regular if minimum value of $H(m)$ is zero. Therefore $H(m)$ the permutation entropy provide a quantitative measure of dynamical complexity of a time series as it refers to its local structures. It may be noted that the permutation entropy depends on the chosen value of m. For $m < 3$, there will be very few distinct states and the foregoing scheme does not work satisfactorily. In the present analysis we have considered sufficiently large time series and chosen $m = 6$ to estimate the complexity measure *MPE*.

For the financial model, Eq. (23), we have simulated the permutation entropy as a function of the scale s for $m = 6$. The simulated results have been shown in **Figure 23a** where we observe a saturation behavior with increased value of the scale factor s. It is also observed that the permutation entropy at any scale s for the interest rate X time series is higher than the investment demand time series Y which

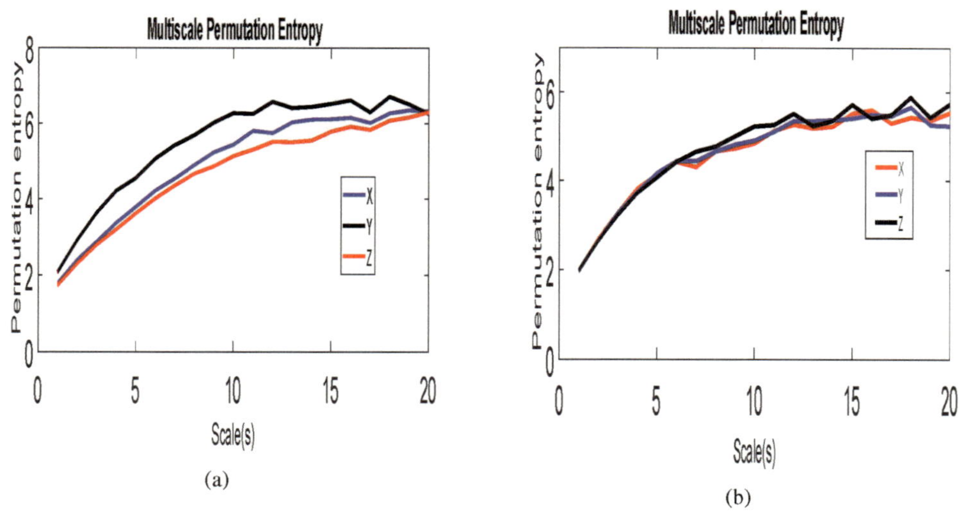

Figure 23.
Multiscale permutation of (a) Financial model time series, (b) Rossler model.

is further higher than that of the price index Z time series. Therefore we may conclude that the complexity measure relation for the considered financial model time series can be expressed as $MPE_X > MPE_Y > MPE_Z$. The behavior of the complexity measure of the considered finance model has been found to be quite similar to that of chaotic Rossler attractor [with parameters $a = 0.15$; $b = 0.20$; $c = 10.0$] (**Figure 23b**).

The simulation results for the multi-scale permutation entropy, *MPE*, presented for the financial model and the Rossler chaotic model exhibit long term correlation of the respective time series of a dynamical variable. Such inference is made in view of the increasing trend of *MPE* with scale factor s for a given m. In case of a standard financial model, the efficacy of such model could be made on comparing the *MPE* trend of resulting simulated time series for interest rate (X), investment demand (Y) and that of price index (Z) with the availability of the real time series data for the corresponding dynamical variables. Finally, we introduce the idea of generation of time series of a nonlinear chaotic dynamical system, say a Lorenz system, using artificial neural network.

7. Time series generation using artificial neural network (ANN)

A large class of different architecture have been used in neural network for various application. Among these application an issue relates the approximation of a nonlinear mapping$f(\mathbf{x})$ with the network $f_{ANN}(\mathbf{x})$, $\mathbf{x} \in \mathbf{R}^{\mathbf{K}}$ where K corresponds to the size of te input. Besides the Radial Basis Function (RBF), a Multi Layer Perceptron (MLP) has been used extensively in function approximation. A MLP neural network comprises an *input layer*, several *hidden layers* and an *output layer* as shown in **Figure 24**.

An *MLP* comprises inputs x_i, $i = 1, 2, \cdots, K$ to the neurons gets multiplied with weights w_{ki} and summed up along with the bias θ_i. The resulting n_i is then acts as an input to the activation function g which could be chosen as a sigmoid function or a

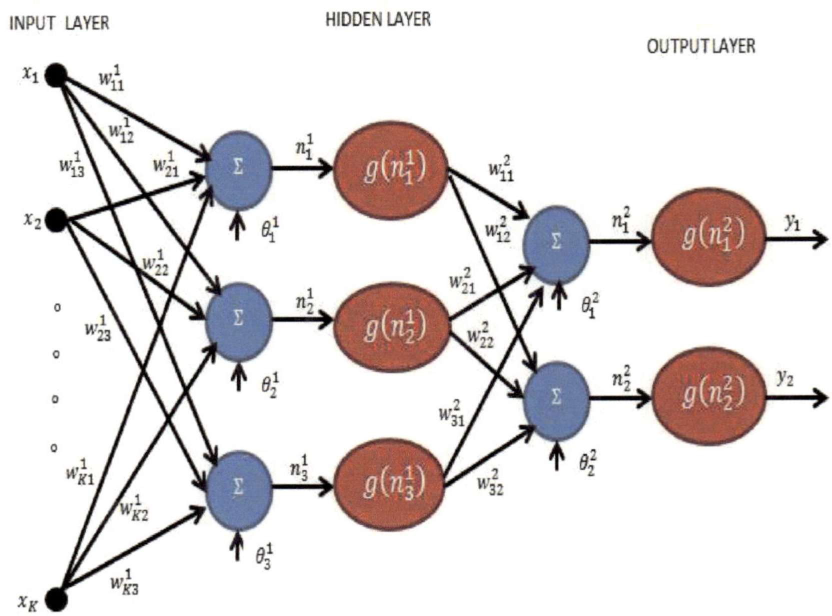

Figure 24.
A multilayer perceptron network.

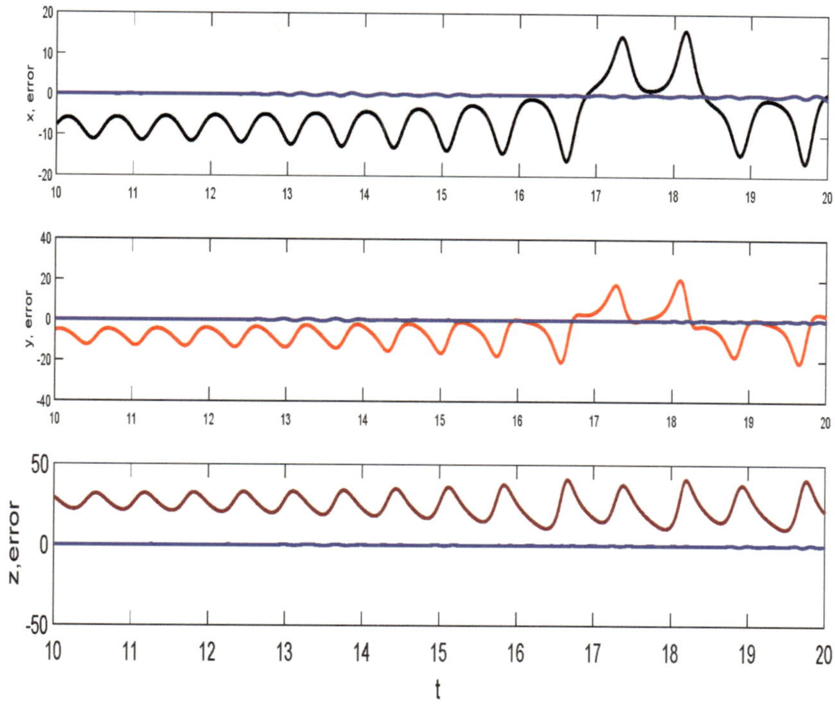

Figure 25.
Neural network generated Lorenz time series (black, red, brown) and respective deviation from input time series (blue).

tanh function. The output at node i is given by $y_i = g\left[\sum_{j=1}^{K} w_{jix_j+\theta_i}\right]$. **Figure 24** illustrates a typical *MLP* network where the output is given by

$$y_i = g\left[\sum_{j=1}^{3} w_{ji}^2\, g\left(\sum_{k=1}^{K} w_{kj}^1 x_k + \theta_j^1\right) + \theta_j^2\right]. \tag{26}$$

Several algorithms are available to determine the network parameters e.g., weights $(w_{ji}^k))$ and biases(θ_j^k). Such algorithms are termed as *teaching* or *learning* algorithms. The basic procedure involving the learning algorithm of an *MLP* network are: (a) Define the network structure, selecting the activation function and initializing the weights and biases, (b) providing the error estimates and number of epochs for training algorithm before running the training algorithm, (c) the output is simulated using input data to the network and compared with the given output, and (d) finally validating the result with independent data.

In this work, using the inputs as x, y and z time series from Lorenz system exhibiting chaotic dynamics and using the *newff*, *train* and *sim MATLAB* commands [31], we simulated each of these time series. In our simulation we take the learning parameters viz., $net.trainParam.show = 50$; $net.trainParam.lr = 0.05$; $net.trainParam.epochs = 1000$; $net.trainParam.goal = 1e^{-3}$, and use 100 neurons and 3 output layers [48]. **Figure 25** shows the *MLP* network generated time series of Lorenz variables and the corresponding deviations from the input time series.

8. Conclusion

In this chapter, we have applied phase portrait, bifurcation diagram, Poincare surface of section, LCEs, correlation dimension, topological entropy and multi-scale permutation entropy method to unravel the complexity of various physical systems e.g., nonlinear forced pendulum, child's swing problem, prey-predator system, periodically stimulated FHN neuron model and nonlinear financial model. Important characterization of transition from regular to chaotic dynamics have been made using the foregoing methods. Finally artificial neural network based on multi-layer perceptron network have been shown to satisfactorily generate the time series of dynamical variable of chaotic system such as Lorenz system.

Author details

Mrinal Kanti Das[1] and Lal Mohan Saha[2*]

1 Institute of Informatics and Communication, University of Delhi South Campus, New Delhi, India

2 IIIMIT, Shiv Nadar University, Gautam Budh Nagar, U.P., India

*Address all correspondence to: lmsaha.msf@gmail.com

References

[1] Stephen Lynch, Dynamical Systems with Applications using Mathematica, 2007, Birkhäuser. Boston, Basel, Berlin.

[2] Patrick T. Tam, A Physicist's Guide to Mathematica, 1997, Academic Press. San Diego, London, New York..

[3] Kinzel, W., and Reents, G., Physics by Computer, 1998, Springer. Berlin, Heidelberg, New York.

[4] Stephen M. Curry, How Children swing, 1976, American Journal of Physics, Vol. 44, P. 924.

[5] Belyakov, A. O., Seyranian, A.O., and Luongo, A., Regular and chaotic dynamics of the swing. ENOC-2008, Saint Petersburg, Russia, June, 30-July, 4, 2008.

[6] Linge, S. O., An assessment of swinger techniques for the playground swing oscillatory motion. Computer Methods in Biomechanics and Biomedical Engineering, 2012, 15:1103–1109.

[7] Libii, J.N., Applying the dynamics of the pendulum to the design of a playground swing. World Transactions on Engineering and Technology Education, 2007, 6: 2,263 – 2,266.

[8] Abrams, P. A.; Ginzburg, L. R., The nature of predation: prey dependent, ratio dependent or neither?. Trends in Ecology & Evolution,2000, 15: 337–341.

[9] Grafton, R.Q., Silva-Echenique, J., Predator–Prey Models: Implications for Management. Atlantic Canada Economics Association Papers, 1994, 23: 61–71.

[10] Swart, J.H., Duffy, K.J., The Stability of a Predator-Prey Model Applied to the Destruction of Trees by Elephants. South African Journal of Science,1987, 18: 56–158.

[11] Collie, J.S., Spencer, P.D.,Modeling Predator-Prey Dynamics in a Fluctuating Environment. Canadian Journal of Fisheries and Aquatic Sciences, 1994, 51: 2665–2672.

[12] Allee, W.C, Bowen E., Studies in animal aggregations: mass protection against colloidal silver among goldfishes. Journal of Experimental Zoology, 1932, 61: 185–207.

[13] Kramer, A.M, Dennis, B., Liebhold, A.M., Drake, J.M., The evidence for Allee effects. Population Ecology, 2009, 51: 341–354.

[14] Berec, L., Angulo, E., Courchamp, F., Multiple Allee effects and population management, Trends in Ecology & Evolution,2007, 22: 185–191.

[15] Wan-Xiong, Yan-Bo-Zhang and Chang-zhong Liu, Analysis of a discrete-time predator-prey system with Allee effect. Ecological Complexity, 2011, 8: 81 – 85.

[16] Ming Zhao and Yunfei Du, Stability of a discrete-time predator-prey system with Allee effect. Nonlinear Analysis and Differential Equations,2016, 4: 225 –233.

[17] Martelli,M., Introduction to Discrete Dynamical Systems and Chaos. John Wiley & Sons, Inc., 1999, New York.

[18] Argentina, M., Coullet, P., and Krinsky, V., Head-on collisions of waves in an excitable Fitzhugh-Nagumo System: a transition from wave annihilation to classical wave behavior, 2000, J.Theor.Biol., 205:47-52.

[19] Marwan N., Romano M.C., Thiel M., Kurths J. Recurrence plots for the analysis of complex systems. Phys. Rep. 2007; 438: 237-329.

[20] Bhardwaj, R., Das, S. Recurrence quantification analysis of a three level

trophic chain model. Heliyon. 2019, 5 (8). https://doi.org/10.1016/j.heliyon.2019.e02182.

[21] Bhattacharjee, A., Das, M.K., Emergent dynamics of spiking neurons with fluctuating threshold,2016, CNSNs, 46:126–134.

[22] Hugh, R.W., Spikes decisions and actions: dynamical foundations of neuroscience, 2005, Oxford University Press Inc., New York.

[23] Izhikevich E.M., Dynamical systems in neuroscience: the geometry of excitability and bursting,2007 London: MIT press.

[24] Mitsunaga, K., Totoki, Y., and Matsuo, T., Firing pattern estimation of biological neuron model by adaptive observer,2007, LNCS vol. 4984, pp. 83-92.

[25] Faghih, R.T., Savla,K., Dahleh, M. A., and Brown, E. N., The fitzhugh-nagumo model: Firing modes with time-varying parameters & parameter estimation,2010, 32nd Annual International Conference of the IEEE EMBS, pp. 4116-4119.

[26] Gong, Y. , Wang, M. , Hou, Z. , Xin, H., Optimal spike coherence and synchronization on complex Hodgkin-Huxley neuron networks. 2005, Chemphyschem, 6:1042.

[27] Chunxiao, H., Ruixue, L., Shuyan, R., Li, Y., Yanqiu, C., Synchronization of coupled chaotic neurons with unknown time delays via adaptive backstepping control. 2013, Res. J. Appl. Sci. Eng. Technol, 5:5509.

[28] Chang-Woo, S. , Sang-Gui, L. , Seunghwan, K., Stochastic excitation of coherent dynamical states of two coupled FitzHugh-Nagumo neurons.2006, J. Korean. Phys. Soc., 48:179.

[29] Kostova,T., Ravindran, R., and Schonbek, M., Fitzhugh-Nagumo revisited: types of bifurcations, periodical forcing and stability regions by a Lyapunov functional, 2004, IJBC, 14:913-925.

[30] Ringqvist, M., Zhou, Y., On existence andnonexistence of limit cycles for Fitzhugh-Nagumo class models, 2006, New directions and applications in control theory, p337-351, Springer, Berlin.

[31] Fraser, A.M., Swinney, H.L. Independent coordinates for strange attractors from mutual information. 1986, Phys.Rev A, 33:1134-1140.

[32] Kennel, M.B., Brown, R., Abarbanel, H.D.I. Determining embedding dimension for phasespace reconstruction using a geometrical construction. 1992, Phys.Rev. A, 45: 3403-3411.

[33] Delage, O., Bourdier, A. Selection of optimal Embedding Parameters Applied to Short and Noisy Time Series from Rossler System. 2017, J.Mod.Phys.8: 1607-1632.

[34] Bhattacharjee, A., Das, M.K., Bhatraju, N., and Yuasa, M., Spike Dynamics of FHN Neuron with Time Varying Parameters, 2013, Proceedings of the World Congress on Engineering 2013 Vol I, WCE 2013, July 3-5, London, U.K.

[35] Gao, Y., Chaos and bifurcation in the space-clamped fitzhugh-nagumo system, 2004, Chaos Solitons & Fractals, 21:943-956.

[36] Chian AC-L. Nonlinear dynamics and chaos in macroeconomics., 2000, Int J Theor Appl Finance, 3:601.

[37] Chian AC-L, Borotto FA, Rempel EL, Rogers C. Attractor merging crisis in chaotic business cycles., 2005, Chaos, Solitons & Fractals 24:869–875.

[38] Chian AC-L, Rempel EL, Rogers C. Complex economic dynamics: chaotic saddle, crisis and intermittency.,2006, Chaos, Solitons & Fractals, 29: 1194–1218.

[39] Ma JH, Chen YS. Study for the bifurcation topological structure and the global complicated character of a kind of nonlinear finance system (I).,2001, Appl Math Mech (English ed.), 22: 1240–1251.

[40] Ma JH, Chen YS. Study for the bifurcation topological structure and the global complicated character of a kind of nonlinear finance system (II).2001, Appl Math Mech (English ed.), 22: 1375–1382.

[41] Chen WC. Nonlinear dynamics and chaos in a fractional-order financial system., 2008, Chaos, Solitons & Fractals,36:1305–14.

[42] Puu T. Nonlinear economic dynamics.,1989, Lecture notes in economics and mathematical systems, vol. 336. Berlin: Springer.

[43] Nonlinear dynamics and heterogeneous interacting agentsThomas L, Reitz S, Samanidou E, editors, 2005, Lecture notes in economics and mathematical systems, vol. 550. Berlin: Springer.

[44] Allen RGD. Mathematical economics. London: Macmillan; 1973.

[45] Chen, W.-C. : Dynamics and control of a financial system with time-delayed feedbacks, 2008, Chaos, Solitons and Fractals, 37:1198–1207.

[46] Jain, A., Das, M. K. :Modeling Complex Behavior of Financial system: Effect of Time- delayed Feedback, 2019, Proceedings of Knowledge Forum, IIT-Channai.

[47] Bandt, C., Pompe, B.: Permutation entropy: a natural complexity measure for time series., 2002, Phys. Rev. Lett. 88:174102.

[48] Koivo, H. N., 2008, Neural Networks: Basics using MATLAB Neural Network Toolbox.

Chapter 4

Invariants for a Dynamical System with Strong Random Perturbations

Elena Karachanskaya

Abstract

In this chapter we consider the invariant method for stochastic system with strong perturbations, and its application to many different tasks related to dynamical systems with invariants. This theory allows constructing the mathematical model (deterministic and stochastic) of actual process if it has invariant functions. These models have a kind of jump-diffusion equations system (stochastic differential Itô equations with a Wiener and a Poisson paths). We show that an invariant function (with probability 1) for stochastic dynamical system under strong perturbations exists. We consider a programmed control with Prob. 1 for stochastic dynamical systems – PSP1. We study the construction of stochastic models with invariant function based on deterministic model with invariant one and show the results of numerical simulation. The concept of a first integral for stochastic differential equation Itô introduce by V. Doobko, and the generalized Itô – Wentzell formula for jump-diffusion function proved us, play the key role for this research.

Keywords: Itô equation, Poisson jump, invariant function, differential equations system construction, stochastic system with invariants, programmed control with probability 1

1. Introduction

Models for actual dynamical processes are based on some restrictions. These restrictions are represented as a conservation law.

The conservation law states that a particular measurable property of an isolated dynamical system does not change as the system evolves over time.

Actual dynamical systems are open, and they are subject to strong external disturbances that violate the laws of conservation for the given system.

Conventionally, deterministic dynamical systems have an invariant function. Doobko[1] V. in [1] proved that stochastic dynamical systems have an invariant function as well. For dynamical system which are described using a system of stochastic differential Itô equations, a first integral – or an invariant function, exists with probability 1 [2–10].

When we know only a conservation law for a dynamical system, and equations which describing this system are unknown, the invariant functions are a good tool for determination of these equations.

[1] Different variant of transliteration of the name: Dubko

Our method differs for other (see, for example, [11]) preliminary in the fact that we construct a system of differential equation with the given first integral under arbitrary initial conditions. Besides, this algorithm is realized as software and it allows us to choose a set of functions for simulation. Moreover, we can construct both a system of stochastic differential equations and a system of deterministic ones.

The goal of this chapter is representation of modern approach to describe of dynamical systems having a set of invariant functions.

This chapter is structured as follows. Firstly, we show that the invariant functions for stochastic systems exist. Then, the generalized Itô – Wentzell formula is represented. It is a differentiated rule for Jump-diffusion function under variables which solves the Jump-diffusion equations system. This rule is basic for the necessary and sufficient conditions for the stochastic first integral (or invariant function with probability 1) for the Jump-diffusion equations system. The next step is the construction of the differential equations system using the given invariant functions. It can be applied for stochastic and nonstochastic cases. The concept of PCP1 (Programmed control with Prob. 1) for stochastic dynamical systems is introduced. Finally, we show an application of the stochastic invariant theory for a transit from deterministic model with invariant to the same stochastic model. Several examples of application of this theory are given and confirmed by results of numerical calculations.

2. Notation and preliminaries

Now we introduce the main concepts which we will use below.

Let $w(t), t \in [0, \infty)$ be a Wiener process or a (standard) Brownian motion, i. e.

- $w(0) = 0$,
- it has stationary, independent increments,
- for every $t > 0$, $w(t)$ has a normal $\mathcal{N}(0, t)$ distribution,
- it has continuous sample paths,
- every trajectory of $w(t)$ is not differentiated for all $t \geq 0$.

A $\nu(t, A)$ is called a Poisson random measure or standard Poisson measure (PM) if it is non-negative integer random variable with the Poisson distribution $\nu(t, A) \sim Poi(t\Pi(A))$, and it has the properties of measure:

- $\nu(t, A)$ is a random variable for every $t \in [0, T]$, $A \in \mathbb{R}^{n'}$,
- $\nu(t, A) \in \mathbb{N} \cup \{0\}$, $\nu(t, \varnothing) = 0$,
- if $A \cap B = \varnothing$, then $\nu(t, A \cup B) = \nu(t, A) + \nu(t, B)$,
- $\mathbf{E}[\nu(t, \mathcal{A})] = t\Pi(A)$,
- if $\#A$ is a number of random events from set A during t, then

$$\mathbf{P}_t(\#A = k) = \frac{(t\Pi(A))^k}{k!} \exp\{-t\Pi(A)\}.$$

$\tilde{\nu}(t, A) = \nu(t, A) - \mathbf{E}[\nu(t, A)]$ is called a centered Poisson measure (CPM).

Let $\mathbf{w}(t) = (w_1(t), \dots, w_m(t))^*$ be an m-dimensional Wiener process, such that the one-dimensional Wiener processes $w_k(t)$ for $k = 1, \dots, m$ is mutually independent.

Take a vector $\gamma \in \Theta$ with values in $\mathbb{R}^{n'}$. Denote by $\nu(\Delta t, \Delta\gamma)$ the PM on $[0, T] \times \mathbb{R}^{n'}$ modeling independent random variables on disjoint intervals and sets. The Wiener processes $w_k(t)$, $k = 1, \dots, m$, and the Poisson measure $\nu([0, T], A)$ defined on the specified space are $\mathcal{F}_t$-measurable and independent of one another.

Consider a random process $x(t)$ with values in $\mathbf{R}^n$, $n \geq 2$, defined by the Equation [12]:

$$dx(t) = A(t)dt + B(t)d\mathbf{w}(t) + \int_{R_\gamma} g(t, \gamma, \mathbf{x})\nu(dt, d\gamma), \tag{1}$$

where $A(t) = \{a_1(t), \dots, a_n(t)\}^*$, $B(t) = \left(b_{j,k}(t)\right)$ is $(n \times k)$ - matrix, and $g(t, \gamma) = \left\{g_1(t, \gamma), \dots, g_n(t, \gamma)\right\}^* \in \mathbb{R}^n$, and $\gamma \in \mathbb{R}^{n'} =: R_\gamma$, while $\mathbf{w}(t)$ is an m-dimensional Wiener process. In general the coefficients $A(t)$, $B(t)$, and $g(t, \gamma)$ are random functions depending also on $\mathbf{x}(t)$. Since the restrictions on these coefficients relate explicitly only to the variables t and γ, we use precisely this notation for the coefficients of (1) instead of writing $A(t, \mathbf{x}(t))$, $(t, \mathbf{x}(t))$, and $g(t, \mathbf{x}(t), \gamma)$.

A system (1) is the stochastic differential Itô equation with Wiener and Poisson perturbations, which named below as a Jump-diffusion Itô equations system (GSDES).

We will consider the dynamical system described using ordinary deterministic differential equations (ODE) system and ordinary stochastic differential Itô equations (SDE) system of different types, taking into account the fact that $\mathbf{x} \in \mathbf{R}^n$, $n \geq 2$.

3. An existence of an invariant function (with Prob.1) for stochastic dynamical system under strong perturbations

Consider the diffusion Itô equation in $\mathbf{R}^3$ with orthogonal random action with respect to the vector of the solution

$$d\mathbf{v}(t) = -\mu\mathbf{v}(t)\,dt + \frac{b}{|\mathbf{v}(t)|}\,[\mathbf{v}(t) \times d\mathbf{w}(t)], \tag{2}$$

where $\mathbf{v} \in \mathbf{R}^3$, $\mathbf{w} \in \mathbf{R}^3$, and $w_i(t)$, $i = 1, 2, 3$ are independent Wiener processes. This equation is a specific form of the Langevin equation.

V. Doobko in [1] showed that the system (2) have an invariant function called a first integral of this system:

$$u(t, \mathbf{v}) = \exp\{2\mu t\}\left(|\mathbf{v}(0)|^2 - \frac{b^2}{\mu}\right).$$

This, in particular, implies that

$$\lim_{t \to \infty} |\mathbf{v}(t)|^2 = \frac{b^2}{\mu},$$

i.e. process $|\mathbf{v}(t)|$ is a nonrandom function and the random process $\mathbf{v}(t)$ itself is generated in a sphere of constant radius $\frac{b}{\sqrt{\mu}}$.

In [4, 5, 10] it is shown that invariant function exists for other stochastic equations of Langevin type. To obtain this result, it is necessary to use the Itô's formula.

4. The generalized Itô – Wentzell formula for jump-diffusion function

The rules for constructing stochastic differentials, e.g., the change rule, are very important in the theory of stochastic random processes. These are Itô's formula [13, 14] for the differential of a nonrandom function of a random process and the Itô – Wentzell[2] formula [15] enabling us to construct the differential of a function which per se is a solution to a stochastic equation. Many articles address the derivation of these formulas for various classes of processes by extending Itô's formula and the Itô – Wentzell formula to a larger class of functions.

The next level is to obtain a new formula for the generalized Itô Equation [14] which involves Wiener and Poisson components. In 2002, V. Doobko presented [7] a generalization of stochastic differentials of random functions satisfying GSDES with CPM based on expressions for the kernels of integral invariants (only the ideas of a possible proof) were sketched in [7]. The result is called" the generalized Itô – Wentzell formula".

In contrast to [7], the generalized Itô – Wentzell formula for the noncentered Poisson measure was represented in [9, 16, 17]. The proof [9] of the generalized Itô – Wentzell formula uses the method of stochastic integral invariants and equations for their kernels. In this case the requirement on the character of the Poisson distribution is only a general restriction, as the knowledge of its explicit form is unnecessary. Other proofs in [16, 17] are based on traditional stochastic analysis and the use of approximations to random functions related to stochastic differential equations by averaging their values at each point.

The generalized Itô – Wentzell formula relying on the kernels of integral invariants [9] requires stricter conditions on the coefficients of all equations under consideration: the existence of second derivatives. The reason is that the kernels of invariants for differential equations exist under certain restrictions on the coefficients.

Since the random function $F(t, \mathbf{x}(t))$ has representation as stochastic diffusion Itô equation with jumps, we can use the generalized Itô – Wentzell formula, proved by us by several methods in accordance with different conditions for the equations coefficients. Now we consider only one case.

We will use the following notation: $\mathcal{C}^s_y$ is the space of functions having continuous derivatives of order s with respect to y, $\mathcal{C}^s_0(y)$ is the space of bounded functions having bounded continuous derivatives of order s with respect to y.

Theorem 1.1 **(generalized Itô – Wentzell formula)**. Consider the real function $F(t,\mathbf{x}) \in \mathcal{C}^{1,2}_{t,x}$, $(t,\mathbf{x}) \in [0,T] \times \mathbb{R}^n$ with generalized stochastic differential of the form

$$d_tF(t,\mathbf{x}) = Q(t,\mathbf{x})dt + \sum_{k=1}^{m} D_k(t,\mathbf{x})dw_k(t) + \int_{R_\gamma} G(t,\mathbf{x},\gamma)\nu(dt,d\gamma) \tag{3}$$

whose coefficients satisfy the conditions:

$$Q(t,\mathbf{x}) \in \mathcal{C}^{1,2}_{t,x}, \quad D_k(t,\mathbf{x}) \in \mathcal{C}^{1,2}_{t,x}, \quad G(t,\mathbf{x},\gamma) \in \mathcal{C}^{1,2,1}_{t,x,\gamma}.$$

[2] Different variants of transliteration of this formula name: Itô – Wentcell, Itô – Venttcel', Itô – Ventzell

If a random process $\mathbf{x}(t)$ obeys (1) and its coefficients satisfy the conditions

$$a_i(t,\mathbf{x}) \in C^{1,1}_{t,x}, \quad b_{ij}(t,\mathbf{x}) \in C^{1,2}_{t,x}, \qquad g_i(t,\mathbf{x},\gamma) \in C^{1,2,1}_{t,x,\gamma}. \tag{4}$$

then the stochastic differential exists and

$$\begin{aligned} d_tF(t,\mathbf{x}(t)) &= Q(t,\mathbf{x}(t))dt + \sum_{k=1}^{m} D_k(t,\mathbf{x}(t))d\mathrm{w}_k + \\ &+ \left[\sum_{i=1}^{n} a_i(t) \frac{\partial F(t,\mathbf{x})}{\partial x_i}\bigg|_{\mathbf{x}=\mathbf{x}(t)} + \frac{1}{2}\sum_{i=1}^{n}\sum_{j=1}^{n}\sum_{k=1}^{m} b_{i,k}(t) b_{j,k}(t) \frac{\partial^2 F(t,\mathbf{x})}{\partial x_i \partial x_j}\bigg|_{\mathbf{x}=\mathbf{x}(t)} + \right. \\ &\left. + \sum_{i=1}^{n} b_{i,k}(t) \frac{\partial D_k(t,\mathbf{x})}{\partial x_i}\bigg|_{\mathbf{x}=\mathbf{x}(t)} \right] dt + \sum_{i=1}^{n}\sum_{k=1}^{m} b_{i,k}(t) \frac{\partial F(t,\mathbf{x})}{\partial x_i}\bigg|_{\mathbf{x}=\mathbf{x}(t)} d\mathrm{w}_k + \\ &+ \int_{R_\gamma} [(F(t,\mathbf{x}(t) + g(t,\gamma)) - F(t,\mathbf{x}(t))]\nu(dt,d\gamma) + \\ &+ \int_{R_\gamma} G(t,\mathbf{x}(t) + g(t,\gamma),\gamma)\nu(dt,d\gamma). \end{aligned} \tag{5}$$

By analogy with the terminology proposed earlier, let us call formula (5) "the generalized Itô – Wentzell formula for the GSDES with PM" (GIWF).

By analogy with the classical Itô and Itô – Wentzell formulas, the generalized Itô – Wentzell formula is promising for various applications. In particular, it helped to obtain equations for the first and stochastic first integrals of the stochastic Itô system [9], equations for the density of stochastic dynamical invariants, Kolmogorov equations for the density of transition probabilities of random processes described by the generalized stochastic Itô differential Equation [8], as well as the construction of program controls with probability 1 for stochastic systems [18, 19].

5. A first integral for GSDES

In the theory of ODE, there are constructed equations to find deterministic functions, first integrals which preserve a constant value with any solutions to the equation. The concept of a first integral plays an important role in theoretical mechanics, for example, to solve inverse problems of mechanics or in constructing controls of dynamical systems.

It turned out that the first integral exists in the theory of stochastic differential equations (SDE) as well. However, there appears an additional classification connected with different interpretations. This gives a first integral for a system of SDE (see [1]), a first direct integral, and a first inverse integral for a system of Itô SDE (see [20]).

Definition 1.1 [1, 3]. Let $\mathbf{x}(t)$ be an n-dimensional random process satisfying a system of Itô SDE

$$dx_i(t) = a_i(t,\mathbf{x}(t))dt + \sum_{k=1}^{m} b_{ik}(t,\mathbf{x}(t))d\mathrm{w}_k(t), \qquad \mathbf{x}(t,\mathbf{x}(0))|_{t=0} = \mathbf{x}(0), \tag{6}$$

whose coefficients satisfy the conditions of the existence and uniqueness of a solution [12]. A nonrandom function $u(t,\mathbf{x}) \in C^{1,2}_{t,x}$ is called a first integral of the system of SDE if it takes a constant value depending only on $\mathbf{x}(0)$ on any trajectory solution to (6) with probability 1:

$$u(t, \mathbf{x}(t, \mathbf{x}(0)) = u(0, \mathbf{x}(0)) \quad \text{almost surely},$$

or, in other words, its stochastic differential is equal to zero: $d_t u(t, \mathbf{x}(t)) = 0$.

Another important notion in the theory of deterministic dynamical systems is given by the notion of an integral invariant introduced by Poincaré [21].

As it turned out, there also exist integral invariants for stochastic dynamical systems [2, 3]. In [7] V. Doobko give the concept of a kernel (=density) of a stochastic integral invariant and, based on it, formulate the notion of a stochastic first integral and a first integral as a deterministic function for GSDES with the centered Poisson measure, which makes it possible to compose a list of first integrals for stochastic differential equations.

Consider a random process $\mathbf{x}(t)$, $\mathbf{x} \in \mathbb{R}^n$, which is a solution to GSDES

$$\begin{aligned} dx_i(t) &= a_i(t, \mathbf{x}(t))dt + b_{ik}(t, \mathbf{x}(t))dw_k(t) + \int_{R_\gamma} g_i(t, \mathbf{x}(t), \gamma)\nu(dt, d\gamma), \\ \mathbf{x}(t) &= x(t, \mathbf{x}(0), \omega)|_{t=0} = \mathbf{x}(0), \qquad i = 1, \dots, n, \qquad t \geq 0, \end{aligned} \tag{7}$$

whose coefficients (in general, random functions) satisfy the conditions of the existence and uniqueness of a solution [12] and the following smoothness conditions:

$$a_i(t, \mathbf{x}) \in C^{1,1}_{t,x}, \quad b_{ij}(t, \mathbf{x}) \in C^{1,2}_{t,x}, \quad g_i(t, \mathbf{x}, \gamma) \in C^{1,2,1}_{t,x,\gamma}. \tag{8}$$

Suppose that $\rho(t, \mathbf{x}, \omega)$ is a random function connected with any deterministic function $f(t, \mathbf{x}) \in \mathfrak{S} \subset C^{1,2}_0(t, \mathbf{x})$ by the relations

$$\int_{\mathbb{R}^n} \rho(t, \mathbf{x}, \omega) f(t, \mathbf{x}) d\hat{\Gamma}(x) = \int_{\mathbb{R}^n} \rho(0, \mathbf{y}) f(t, \mathbf{x}(t, \mathbf{y})) d\hat{\Gamma}(\mathbf{y}) \tag{9}$$

$$\int_{\mathbb{R}^n} \rho(0, \mathbf{x}) d\hat{\Gamma}(\mathbf{x}) = 1, \tag{10}$$

$$\lim_{|\mathbf{x}| \to \infty} \rho(0, \mathbf{x}, \omega) = \lim_{|\mathbf{x}| \to \infty} \rho(0, \mathbf{x}) = 0, \qquad d\hat{\Gamma}(\mathbf{x}) = \prod_{i=1}^{n} dx_i, \tag{11}$$

where $\mathbf{y} := \mathbf{x}(0)$, and $\mathbf{x}(t, \mathbf{y})$ is a solution to (7), and ω is a random event.

In the particular case when $f(t, \mathbf{x}) = 1$, conditions (9) and (10) imply that

$$\int_{\mathbb{R}^n} \rho(t, \mathbf{x}, \omega) d\hat{\Gamma}(\mathbf{x}) = \int_{\mathbb{R}^n} \rho(0, \mathbf{y}) d\hat{\Gamma}(\mathbf{y}) = 1, \tag{12}$$

i.e., for the random function $\rho(t, \mathbf{x}, \omega)$, there exists a nonrandom functional preserving a constant value:

$$\int_{\mathbb{R}^n} \rho(t, \mathbf{x}, \omega) d\hat{\Gamma}(\mathbf{x}) = 1. \tag{13}$$

Then, with conditions (10) and (11), Eq. (9) can be regarded as a stochastic integral invariant, and the function (t, x) can be viewed as its density.

Definition 1.2 [3]. A nonnegative random function $\rho(t, \mathbf{x}, \omega)$ is referred to as a stochastic kernel or the stochastic density of a stochastic integral invariant (of nth order) if conditions (9), (10), and (11) are held.

Note that a substantial difference which made it possible to consider the invariance of the random volume on the basis of a kernel of an integral operator in [3, 7], is that (9) contains a functional factor. Thus, the notion of a kernel of an integral invariant [3] for a system of ordinary differential equations can be regarded as a particular case by taking $f(t, \mathbf{x}) = 1$ and excluding from (7) the randomness determined by the Wiener and Poisson processes.

Using the GIWF (5), we obtain equation for the stochastic kernel function [9].

$$d_t\rho(t,\mathbf{x},\omega) = -\frac{\partial\rho(t,\mathbf{x},\omega)b_{ik}(t,\mathbf{x})}{\partial x_i}dw_k(t) + \Big(-\frac{\partial(\rho(t,\mathbf{x},\omega)a_i(t,\mathbf{x}))}{\partial x_i} + \\ +\frac{1}{2}\frac{\partial^2(\rho(t,\mathbf{x},\omega)b_{ik}(t,\mathbf{x})b_{jk}(t,\mathbf{x}))}{\partial x_i\partial x_j}\Big)dt + \\ +\int_{R_\gamma}\left[\rho\left(t,\mathbf{x}-g\left(t,\mathbf{x}^{-1}(t,\mathbf{x},\gamma),\gamma,\omega\right)\right)\cdot\mathcal{J}\left(\mathbf{x}^{-1}(t,\mathbf{x},\gamma)\right)-\rho(t,\mathbf{x},\omega)\right]\nu(dt,d\gamma), \tag{14}$$

under restrictions

$$\rho(t,\mathbf{x},\omega)|_{t=0} = \rho(0,\mathbf{x},\omega) = \rho(0,\mathbf{x}) \in C_0^2(\mathbf{x}),$$
$$\lim_{|x|\to\infty}\rho(0,\mathbf{x},\omega) = \lim_{|\mathbf{x}|\to\infty}\rho(0,\mathbf{x}) = 0, \quad \lim_{|\mathbf{x}|\to\infty}\frac{\partial\rho(0,\mathbf{x},\omega)}{\partial x_i} = \lim_{|\mathbf{x}|\to\infty}\frac{\partial\rho(0,\mathbf{x})}{\partial x_i} = 0.$$

This result plays a major role in obtaining of equation for the stochastic first integral.

6. Necessary and sufficient conditions for the stochastic first integral

Lemma 1.1. If $\rho(t, \mathbf{x}, \omega)$ is a stochastic kernel of an integral invariant of n th order of a stochastic process $\mathbf{x}(t)$ starting from a point $\mathbf{x}(0)$ then, for every t, it satisfies the equality

$$\rho(t,\mathbf{x}(t,\mathbf{x}(0)),\omega)\mathcal{J}(t,\mathbf{x}(0),\omega) = \rho(0,\mathbf{x}(0)),$$

where $\mathcal{J}(t, \mathbf{x}(0), \omega)$ is the Jacobian of transition from $\mathbf{x}(t)$ to $\mathbf{x}(0)$.

Definition 1.3 A set of kernels of integral invariants of nth order is called complete if any other function that is the kernel of this integral invariant can be presented as a function of the elements of this set.

In [9] it is shown that a system of GSDE (7) whose coefficients satisfy the conditions in (8), has a complete set of kernels consisting of $(n + 1)$ functions.

Suppose that $\rho_l(t, \mathbf{x}, \omega) \neq 0$, $l = 1, \ldots, m$, $m \leq n + 1$ are kernels of the integral invariant (9). Lemma 1.1 implies that, for any $l \neq n + 1$, the ratio $\frac{\rho_l(t,\mathbf{x}(t,\mathbf{y}),\omega)}{\rho_{n+1}(t,\mathbf{x}(t,\mathbf{y}),\omega)}$ is a constant depending only on the initial condition $\mathbf{x}(0) = \mathbf{y}$ for every solution $\mathbf{x}(t)$ to the GSDE (7) because

$$\frac{\rho_s(t,\mathbf{x}(t,\mathbf{y}),\omega)}{\rho_{n+1}(t,\mathbf{x}(t,\mathbf{y}),\omega)} = \frac{\rho_s(0,\mathbf{y})}{\rho_{n+1}(0,\mathbf{y})}. \tag{15}$$

Since for some realization ω_1 we have

$$u(t,\mathbf{x}(t,\mathbf{x}(0))) \equiv \frac{\rho_l(t,\mathbf{x}(t,\mathbf{x}(0)),\omega_1)}{\rho_s(t,\mathbf{x}(t,\mathbf{x}(0)),\omega_1)} = \frac{\rho_l(0,\mathbf{x}(0))}{\rho_s(0,\mathbf{x}(0))} \equiv u(0,\mathbf{x}(0)),$$

and it means, that $d_t u(t,\mathbf{x}(t)) = 0$.

Definition 1.4 A random function $u(t,\mathbf{x},\omega)$ defined on the same probability space as a solution to (7) is referred to as a stochastic first integral of the system (7) of Itô Ë† GSDE with NCM if the following condition holds with probability 1:

$$u(t,\mathbf{x}(t,\mathbf{x}(0),\omega)) = u(0,\mathbf{x}(0)) \quad \text{almost surely}$$

for every solution $\mathbf{x}(t,\mathbf{x}(0),\omega)$ to (7).

For practical purposes, for example, to construct program controls for a dynamical system under strong random perturbations, the presence of a concrete realization is important, i.e., the parameter ω is absent in what follows. In this connection, we introduce one more notion.

Definition 1.5 A nonrandom function $u(t,\mathbf{x})$ is called a first integral of the system of GSDE (7) if it preserves a constant value with probability 1 for every realization of a random process $\mathbf{x}(t)$ that is a solution to this system:

$$u(t,\mathbf{x}(t,\mathbf{x}(0))) = u(0,\mathbf{x}(0)) \quad \text{almost surely.}$$

Thus, a stochastic first integral includes all trajectories (or realizations) of the random process while the first integral is related to one realization.

Construct an equation for $u(t,\mathbf{x},\omega)$ using the relation

$$\ln u_s(t,\mathbf{x},\omega) = \ln \rho_s(t,\mathbf{x},\omega) - \ln \rho_l(t,\mathbf{x},\omega), \tag{16}$$

as a result of assertion (15). Let us differentiate $\ln \rho(t,\mathbf{x})$ (omit ω) using generalized Itô – Wentzell formula:

$$\begin{aligned} d_t \ln \rho(t,\mathbf{x}) &= \frac{1}{\rho(t,\mathbf{x})}\tilde{d}_t\rho(t,\mathbf{x}) - \frac{1}{2\rho^2(t,\mathbf{x})}\left(-\frac{\partial(\rho(t,\mathbf{x})b_{ik}(t,\mathbf{x}))}{\partial x_i}\right)^2 dt + \\ &+\int_{R_\gamma} \left[\ln\left\{\rho_s(t,\mathbf{x}-g(t,\mathbf{x}(t,\mathbf{y}),\gamma),\gamma)\mathcal{J}\left(\mathbf{x}^{-1}(t,\mathbf{x},\gamma)\right)\right\} - \ln \rho_s(t,\mathbf{x})\right]\nu(dt,d\gamma), \end{aligned} \tag{17}$$

where $\tilde{d}_t\rho(t,\mathbf{x})$ is the right side of Eq.(14) without the integral expression. Having written down the equations for $\ln \rho_s(t,\mathbf{x})$ and $\ln \rho_l(t,\mathbf{x})$, and taking into account this result and Eq.(16), we obtain:

$$\begin{aligned} d_t u(t,\mathbf{x},\omega) &= \left[-a_i(t,x)\frac{\partial u(t,x,\omega)}{\partial x_i} + \frac{1}{2} b_{ik}(t,x) b_{jk}(t,x)\frac{\partial^2 u(t,x,\omega)}{\partial x_i \partial x_j} - \right. \\ &\left. -b_{ik}(t,\mathbf{x})\frac{\partial}{\partial x_i}\left(b_{jk}(t,\mathbf{x})\frac{\partial u(t,\mathbf{x},\omega)}{\partial x_j}\right)\right]dt - b_{ik}(t,\mathbf{x})\frac{\partial u(t,\mathbf{x},\omega)}{\partial x_i} dw_k(t) + \\ &+\int_{R_\gamma} \left[u\left(t,\mathbf{x}-g\left(t,\mathbf{x}^{-1}(t,\mathbf{x},\gamma),\gamma\right),\omega\right) - u(t,\mathbf{x},\omega)\right]\nu(dt,d\gamma), \end{aligned} \tag{18}$$

which means that a stochastic first integral $u(t,\mathbf{x},\omega)$ of the Itô generalized Eq. (7) is a solution to the GSDE (18).

For a first integral which is a nonrandom function of one realization, the differential is also defined by an equation of the form of (18).

Theorem 1.2 Let $\mathbf{x}(t)$ be a solution to the GSDES (7) with conditions (8). A nonrandom function $u(t,\mathbf{x}) \in C^{1,2}_{t,x}$ is a first integral of system (7) if and only if it satisfies the conditions:

1. $\frac{\partial u(t,\mathbf{x})}{\partial t} + \frac{\partial u(t,\mathbf{x})}{\partial x_i}\left[a_i(t,\mathbf{x}) - \frac{1}{2} b_{jk}\left(t,\mathbf{x}\right)\frac{\partial b_{ik}(t,\mathbf{x})}{\partial x_j}\right] = 0,$

2. $b_{ik}(t,\mathbf{x})\frac{\partial u(t,\mathbf{x})}{\partial x_i} = 0$, for all $k = \overline{1,m}$,

3. $u(t,\mathbf{x}) - u(t,\mathbf{x} + g(t,\mathbf{x},\gamma)) = 0$ for any $\gamma \in R_\gamma$ in the entire domain of definition of the process.

Theorem (6) allows us to obtain a method for construction of differential equations systems on the basis of the given set of invariant functions.

7. Construction of the differential equations system using the given invariant functions

The concept of a first integral for a system of stochastic differential equations plays a key role in our theory. In this section, we will use a set of first integrals for the construction of a system of differential equations.

Let us write Eq. (7) in matrix form:

$$dX(t) = A(t,X(t))dt + B(t,X(t))dw(t) + \int_{R_\gamma} \Theta(t,X(t),\gamma)\nu(dt,d\gamma)$$
$$X(0) = x_0, \qquad t \geq 0. \tag{19}$$

Theorem 1.3 [22]. Let $X(t)$ be a solution of the Eq. (19) and let a nonrandom function $s(t,x)$ be continuous together with its first-order partial derivatives with respect to all its variables. Assume the set $\left\{\vec{e}_o, \vec{e}_1, \dots, \vec{e}_n\right\}$ defines an orthogonal basis in $\mathbf{R}_+ \times \mathbf{R}^n$. If function $s(t,x)$ is a first integral for the system (19), then the coefficients of Eq. (19) and the function $s(t,x)$ together are related by the conditions:

1. Functions $B_k(t,x) = \sum_{i=1}^{n} b_{ik}(t,x)\vec{e}_i$ $(k = \{1, \dots, m\})$, which determine columns of the matrix $B(t,x)$, belong to a set

$$B_k(t,x) \in \left\{ q_{oo}(t,x) \cdot \det \begin{bmatrix} \vec{e}_1 & \dots & \vec{e}_n \\ \frac{\partial s(t,x)}{\partial x_1} & \dots & \frac{\partial s(t,x)}{\partial x_n} \\ f_{31} & \dots & f_{3n} \\ \dots & \dots & \dots \\ f_{n1} & \dots & f_{nn} \end{bmatrix} \right\}, \tag{20}$$

where $q_{oo}(t,x)$ is an arbitrary nonvanishing function,

2. Coefficient $A(t,x)$ belongs to a set of functions defined by

$$A(t,x) \in \left\{ R(t,x) + \frac{1}{2}\sum_{k=1}^{n}\left[\frac{\partial B_k(t,x)}{\partial x}\right] \cdot B_k(t,x)\right\}, \tag{21}$$

where a column matrix $R(t,x)$ with components $r_i(t,x), i = \{1, \dots, n\}$, is defined as follows:

$$C^{-1}(t,x) \cdot \det H(t,x) = \vec{e}_o + \sum_{i=1}^{n} r_i(t,x)\vec{e}_i,$$

$C(t,x)$ is an algebraic adjunct of the element $\vec{e}_o$ of a matrix $H(t,x)$ and $\det C(t,x) \neq 0$, a matrix $H(t,x)$ is defined as

$$H(t,x) = \begin{bmatrix} \vec{e}_o & \vec{e}_1 & \dots & \vec{e}_n \\ \frac{\partial s(t,x)}{\partial t} & \frac{\partial s(t,x)}{\partial x_1} & \dots & \frac{\partial s(t,x)}{\partial x_n} \\ h_{30} & h_{31} & \dots & h_{3n} \\ \dots & \dots & \dots & \dots \\ h_{n+1,0} & h_{n+1,1} & \dots & h_{n+1,n} \end{bmatrix}, \tag{22}$$

and $\left[\frac{\partial B_k(t,x)}{\partial x}\right]$ is a Jacobi matrix for function $B_k(t,x)$,

3. Coefficient $\Theta(t,X,\gamma) = \sum_{i=1}^{n} \gamma_i(t,x,\gamma)\vec{e}_i$, related to Poisson measure, is defined by the representation $\Theta(t,x,\gamma) = y(t,x,\gamma) - x$, where $y(t,x,\gamma)$ is a solution of the differential equations system

$$\frac{\partial y(\cdot,\gamma)}{\partial \gamma} = \det \begin{bmatrix} \vec{e}_1 & \vec{e}_2 & \dots & \vec{e}_n \\ \frac{\partial s(t,y(\cdot,\gamma))}{\partial y_1} & \frac{\partial s(t,y(\cdot,\gamma))}{\partial y_2} & \dots & \frac{\partial s(t,y(\cdot,\gamma))}{\partial y_n} \\ \varphi_{31}(t,y(\cdot,\gamma)) & \varphi_{32}(t,y(\cdot,\gamma)) & \dots & \varphi_{3n}(t,y(\cdot,\gamma)) \\ \dots & \dots & \dots & \dots \\ \varphi_{n1}(t,y(\cdot,\gamma)) & \varphi_{n2}(t,y(\cdot,\gamma)) & \dots & \varphi_{nn}(t,y(\cdot,\gamma)) \end{bmatrix}. \tag{23}$$

This solution satisfies the initial condition: $y(t,x,\gamma)|_{\gamma=0} = x$.

The arbitrary functions $f_{ij} = f_{ij}(t,x)$, $h_{ij} = h_{ij}(t,x)$, and $\varphi_{ij} = \varphi_{ij}(t,y(\cdot,\gamma))$ are defined by the equalities $f_{ij}(t,x) = \frac{\partial f_i(t,x)}{\partial x_j}$, $h_{ij}(t,x) = \frac{\partial h_i(t,x)}{\partial x_j}$, and $\varphi_{ij}(t,y(\cdot,\gamma)) = \frac{\partial \varphi_i(t,y(\cdot,\gamma))}{\partial y_j}$. Sets of functions $\{\varphi_i(t,y(\cdot,\gamma))\}$ and the function $g(t,x)$ together form a class of independent functions.

Using this theorem, we can to construct SDE system of different types and ODE system. Choice of arbitrary functions allows us to construct a set of differential equations systems with the given invariant functions. Theorem (7) allows us to introduce a concept of Programmed control with probability 1 for stochastic dynamical system.

8. Programmed control with Prob. 1 for stochastic dynamical systems

Definition 1.6 [18, 19]. A PCP1 is called a control of stochastic system which allows the preservation with probability 1 of a constant value for the same function which depends on this systems position for time periods of any length T.

Let us consider the stochastic nonlinear jump of diffusion equations system:

$$dX(t) = (P(t,X(t)) + Z(t,X(t)) \cdot u(t,X(t)))dt + (B(t,X(t)) + K(t,X(t)))dw(t) + \\ + \int_{R_\gamma} (L(t,X(t),\gamma) + \Lambda(t,X(t)))\nu(dt,d\gamma), \tag{24}$$

where $P(\cdot)$, $Z(\cdot)$ are given matrix functions and $B(\cdot)$, $L(\cdot)$ are the functions that may either be known or not. For such systems we construct a unit of programmed control $\{u(t, X(t)), K(t, X(t)), M(t, X(t))\}$ which allows the system (24) to be on the given manifold $\{u(t, X(t))\} = \{u(0, x_0)\}$ with Prob. 1 (PCP1) for each $t \in [0, T]$, $T \leq \infty$.

Suppose that the nonrandom function $s(t, X(t))$ is the first integral for the same stochastic dynamical system. The PCP1 $\{u(t, X(t)), K(t, X(t)), M(t, X(t))\}$ is the solution for the algebraic system of linear equations.

Theorem 1.4 Let a controlled dynamical system be subjected to Brownian perturbations and Poisson jumps. The unit of PCP1 $\{u(t, X(t)), K(t, X(t)), M(t, X(t))\}$, allowing this system to remain with probability 1 on the dynamically structured integral mfd $s(t, X(t, x_o), \omega) = s(0, x_o)$, is a solution of the linear equations system (with respect to functions $\mathbf{u}(t, \mathbf{x}(t)))$, $K(t, X(t))$, $M(t, X(t))$ which consists of Eq. (19) and Eq. (24). The coefficients of the Eq. (19), (and the coefficients of the Eq. (24) respectively) are determined by the theorem 7. The response to the random action is defined completely.

We show how the stochastic invariants theory can be applied to solve different tasks.

9. Stochastic models with invariant function which are based on deterministic model with invariant one

In this section we consider a few examples for application of the theory above to modeling actual random processes with invariants [23]. Firstly, we consider an example of construction of a differential equation system with the given invariant. Secondly, we study a general scheme for the PCP1 determination. And finally, we show the possibility of construction of stochastic analogues for classical models described by a differential equations system with an invariant function. The suggested method of stochastization is based on both the concept of the first integral for a stochastic differentialItô equations system (SDE) and the theorem for construction of the SDE system using its first integral.

9.1 Construction of a differential equations system

It is necessary to construct a differential equations system for $\mathbf{X} \in \mathbf{R}^3$, $t \geq 0$ such that the equality

$$X(t) - Y_1^2(t) + Y_2(t) + e^t = 0 \tag{25}$$

is satisfied with Prob.1. The equality (25) means that the differential equations system has a first integral $s(t, X(t), Y_1(t), Y_2(t)) = X(t) - Y_1^2(t) + Y_2(t) + e^t$ with initial condition $(0, 1, 0)^*$:

$$s(t, X(t), Y_1(t), Y_2(t)) \equiv X(t) - Y_1^2(t) + Y_2(t) + e^t = s(0, x(0), Y_1(0), Y_2(0)).$$

We have

$$B_k(\cdot) = q_{00}(\cdot)\det\begin{bmatrix} \vec{e}_1 & \vec{e}_2 & \vec{e}_3 \\ 1 & -2y_1 & 1 \\ f_1(\cdot) & f_2(\cdot) & f_3(\cdot) \end{bmatrix} = q_{00}(\cdot)\begin{bmatrix} -2y_1 f_3(\cdot) - f_2(\cdot) \\ -f_3(\cdot) + f_1(\cdot) \\ f_2(\cdot) + 2y_1 f_1(\cdot) \end{bmatrix} = \begin{bmatrix} b_{1k}(\cdot) \\ b_{2k}(\cdot) \\ b_{3k}(\cdot) \end{bmatrix}.$$

Therefore,

$$(B_k(\cdot),\nabla_z)B_k(\cdot)=q_{00}^2(\cdot)\begin{bmatrix}\frac{\partial(-2Y_1f_3(\cdot)-f_2(\cdot))}{\partial x} & \frac{\partial(-2Y_1f_3(\cdot)-f_2(\cdot))}{\partial Y_1} & \frac{\partial(-2Y_1f_3(\cdot)-f_2(\cdot))}{\partial Y_2}\\ \frac{\partial(-f_3(\cdot)+f_1(\cdot))}{\partial x} & \frac{\partial(-f_3(\cdot)+f_1(\cdot))}{\partial Y_1} & \frac{\partial(-f_3(\cdot)+f_1(\cdot))}{\partial Y_2}\\ \frac{\partial(f_2(\cdot)+2Y_1f_1(\cdot))}{\partial x} & \frac{\partial(f_2(\cdot)+2Y_1f_1(\cdot))}{\partial Y_1} & \frac{\partial(f_2(\cdot)+2Y_1f_1(\cdot))}{\partial Y_2}\end{bmatrix}\times$$

$$\times\begin{bmatrix}-2Y_1(t)f_3(\cdot)-f_2(\cdot)\\ -f_3(\cdot)+f_1(\cdot)\\ f_2(\cdot)+2Y_1(t)f_1(\cdot)\end{bmatrix}=\begin{bmatrix}p_1(\cdot)\\ p_2(\cdot)\\ p_3(\cdot)\end{bmatrix}.$$

Thus the new drift coefficients are

$$\begin{aligned}A_1(\cdot)&=\frac{e^t(h_2f_3-f_2h_3)+2Y_1(h_0f_3-f_0h_3)+h_0f_2-f_0h_2}{f_2h_3-h_2f_3+f_1h_2-h_1g_2-2Y_1(h_1f_3-f_1h_3)}+\frac{p_1}{2},\\ A_2(\cdot)&=\frac{e^t(h_3f_1-f_3h_1)+h_0f_3-f_0h_3+h_0f_1-f_0h_1}{f_2h_3-h_2f_3+f_1h_2-h_1g_2-2Y_1(h_1f_3-f_1h_3)}+\frac{p_2}{2},\\ A_3(\cdot)&=\frac{e^t(h_2f_1-f_2h_1)-2Y_1(h_0f_1-f_0h_1)+h_0f_2-f_0h_2}{f_2h_3-h_2f_3+f_1h_2-h_1g_2-2Y_1(h_1f_3-f_1h_3)}+\frac{p_3}{2}.\end{aligned} \tag{26}$$

According to term 3 of Theorem 1.3, we will determine a coefficient for Poisson measure. Now we rename variables: $Z\equiv(Z_1,Z_2,Z_3)^*:=(X,Y_1,Y_2)^*$. Then, we have:

$$\begin{gathered}u(t,Z)=Z_1-Z_2^2+Z_3+e^t,\\ u(t,Z)-u(t,Z+g(t,Z,\gamma))\equiv u(t,Z)-u(t,V)=0,\\ V=Z+g(t,Z,\gamma),\\ g(t,Z,\gamma)=V(t,Z,\gamma)-Z,\end{gathered}$$

where a function $V(t,Z,\gamma)$ solves the a differential equations system:

$$\frac{\partial V(\cdot,\gamma)}{\partial\gamma}=\det\begin{bmatrix}\vec{e}_1 & \vec{e}_2 & \vec{e}_3\\ 1 & -2Z_2 & 1\\ \varphi_1(\cdot,\gamma) & \varphi_2(\cdot,\gamma) & \varphi_3(\cdot,\gamma)\end{bmatrix}=\begin{bmatrix}-2Z_2\varphi_3(\cdot,\gamma)-\varphi_2(\cdot,\gamma)\\ \varphi_1(\cdot,\gamma)-\varphi_3(\cdot,\gamma)\\ \varphi_2(\cdot,\gamma)+2Z_2\varphi_1(\cdot,\gamma),\end{bmatrix},$$

and satisfies the initial condition $V(\cdot,0)=Z$. Then, we determine functions $g_1(\cdot,\gamma),g_2(\cdot,\gamma),g_3(\cdot,\gamma)$.

Assume, that $\varphi_1(\cdot,\gamma)=\frac{1}{\gamma+1}$, $\varphi_2(\cdot,\gamma)=2\gamma$, $\varphi_3(\cdot,\gamma)=1$. Then, we get:

$$\begin{aligned}g_1(t,X(t),Y(t,),\gamma)&=-2Y_1(t)\gamma-\gamma^2-x(t),\\ g_2(t,X(t),Y(t,),\gamma)&=\ln|\gamma+1|-\gamma+1-Y_1(t),\\ g_3(t,X(t),Y(t,),\gamma)&=-\gamma^2+2Y_1\ln|\gamma+1|-Y_2.\end{aligned}$$

Finally, we have constructed three variants of differential equations system:

1. deterministic differential equations system:

$$\begin{cases} \dfrac{dX(t)}{dt} = \dfrac{e^t(h_2f_3 - f_2h_3) + 2Y_1(h_0f_3 - f_0h_3) + h_0f_2 - f_0h_2}{f_2h_3 - h_2f_3 + f_1h_2 - h_1g_2 - 2Y_1(h_1f_3 - f_1h_3)} \\ \dfrac{dY_1(t)}{dt} = \dfrac{e^t(h_3f_1 - f_3h_1) + h_0f_3 - f_0h_3 + h_0f_1 - f_0h_1}{f_2h_3 - h_2f_3 + f_1h_2 - h_1g_2 - 2Y_1(h_1f_3 - f_1h_3)} \\ \dfrac{dY_2(t)}{dt} = \dfrac{e^t(h_2f_1 - f_2h_1) - 2Y_1(h_0f_1 - f_0h_1) + h_0f_2 - f_0h_2}{f_2h_3 - h_2f_3 + f_1h_2 - h_1g_2 - 2Y_1(h_1f_3 - f_1h_3)}. \end{cases}$$

2. stochastic differential equations system (Itó diffusion equations):

$$\begin{cases} dX(t) = \left[\dfrac{e^t(h_2f_3 - f_2h_3) + 2Y_1(h_0f_3 - f_0h_3) + h_0f_2 - f_0h_2}{f_2h_3 - h_2f_3 + f_1h_2 - h_1g_2 - 2Y_1(h_1f_3 - f_1h_3)} + \dfrac{1}{2}(-2Y_1(t)f_3(\cdot) - f_2(\cdot))\right]dt + \\ \qquad + (-2y_1f_3(\cdot) - f_2(\cdot))\ dw_1(t) \\ dY_1(t) = \left[\dfrac{e^t(h_3f_1 - f_3h_1) + h_0f_3 - f_0h_3 + h_0f_1 - f_0h_1}{f_2h_3 - h_2f_3 + f_1h_2 - h_1g_2 - 2Y_1(h_1f_3 - f_1h_3)} + \dfrac{1}{2}(-f_3(\cdot) + f_1(\cdot))\right]dt + \\ \qquad + (-f_3(\cdot) + f_1(\cdot))\ dw_2(t) \\ dY_2(t) = \left[\dfrac{e^t(h_2f_1 - f_2h_1) - 2Y_1(h_0f_1 - f_0h_1) + h_0f_2 - f_0h_2}{f_2h_3 - h_2f_3 + f_1h_2 - h_1g_2 - 2Y_1(h_1f_3 - f_1h_3)} + \dfrac{1}{2}(f_2(\cdot) + 2Y_1(t)f_1(\cdot))\right]dt + \\ \qquad + (f_2(\cdot) + 2y_1f_1(\cdot))\ dw_2(t) \end{cases}.$$

3. stochastic differential equations system (jump-diffusion Itó equations):

$$\begin{cases} dX(t) = \left[\dfrac{e^t(h_2f_3 - f_2h_3) + 2Y_1(h_0f_3 - f_0h_3) + h_0f_2 - f_0h_2}{f_2h_3 - h_2f_3 + f_1h_2 - h_1g_2 - 2Y_1(h_1f_3 - f_1h_3)} + \dfrac{1}{2}(-2Y_1(t)f_3(\cdot) - f_2(\cdot))\right]dt + \\ \qquad + (-2y_1f_3(\cdot) - f_2(\cdot))\ dw_1(t) + \int_{R_\gamma}[-2Y_1(t)\gamma - \gamma^2 - x(t)]\nu(dt, d\gamma), \\ dY_1(t) = \left[\dfrac{e^t(h_3f_1 - f_3h_1) + h_0f_3 - f_0h_3 + h_0f_1 - f_0h_1}{f_2h_3 - h_2f_3 + f_1h_2 - h_1g_2 - 2Y_1(h_1f_3 - f_1h_3)} + \dfrac{1}{2}(-f_3(\cdot) + f_1(\cdot))\right]dt + \\ \qquad + (-f_3(\cdot) + f_1(\cdot))\ dw_2(t) + \int_{R_\gamma}[\ln|\gamma + 1| - \gamma + 1 - Y_1(t)]\nu(dt, d\gamma), \\ dY_2(t) = \left[\dfrac{e^t(h_2f_1 - f_2h_1) - 2Y_1(h_0f_1 - f_0h_1) + h_0f_2 - f_0h_2}{f_2h_3 - h_2f_3 + f_1h_2 - h_1g_2 - 2Y_1(h_1f_3 - f_1h_3)} + \dfrac{1}{2}(f_2(\cdot) + 2Y_1(t)f_1(\cdot))\right]dt + \\ \qquad + (f_2(\cdot) + 2y_1f_1(\cdot))\ dw_2(t) + \int_{R_\gamma}[-\gamma^2 + 2Y_1\ln|\gamma + 1| - Y_2]\nu(dt, d\gamma). \end{cases} \tag{27}$$

We choose the functions $q_{00}(\cdot)$, $f_i(\cdot)$ and $h_i(\cdot)$, $i = 1, 2, 3$, in accordance with the restriction of the task and taking into account the utility for modeling.

9.2 Transit from deterministic model with invariant to the same stochastic model

Now we describe a general scheme for application of the theory above.

The suggested method of stochastization is based on both the concept of the first integral for a stochastic differentialItô equations system (SDE) and the theorem for construction of the SDE system using its first integral.

Let us consider a classical model

$$\begin{cases} dy_1(t) = F_1(t, \mathbf{y}(t))dt, \\ dy_2(t) = F_2(t, \mathbf{y}(t))dt, \\ dy_3(t) = F_3(t, \mathbf{y}(t))dt, \end{cases} \tag{28}$$

with an invariant $u(t, \mathbf{y})$.

Then we construct the GSDE system, taking into account the equality $u(t, \mathbf{x}(t)) = u(0, \mathbf{x}(0)) = C$:

$$\begin{cases} dx_1(t) = a_1(t, \mathbf{x}(t))dt + b_1(t, \mathbf{x}(t))d\mathrm{w}(t) + \int g_1(t, \mathbf{x}(t), \gamma)\nu(dt, d\gamma), \\ dx_2(t) = a_2(t, \mathbf{x}(t))dt + b_2(t, \mathbf{x}(t))d\mathrm{w}(t) + \int g_2(t, \mathbf{x}(t), \gamma)\nu(dt, d\gamma), \\ dx_3(t) = a_3(t, \mathbf{x}(t))dt + b_3(t, \mathbf{x}(t))d\mathrm{w}(t) + \int g_3(t, \mathbf{x}(t), \gamma)\nu(dt, d\gamma). \end{cases} \tag{29}$$

Hence, the stochastic model has a representation

$$\begin{cases} dy_1(t) = a_1(t, \mathbf{y}(t))dt + b_1(t, \mathbf{y}(t))d\mathrm{w}(t) + \int g_1(t, \mathbf{y}(t), \gamma)\nu(dt, d\gamma), \\ dy_2(t) = a_2(t, \mathbf{y}(t))dt + b_2(t, \mathbf{y}(t))d\mathrm{w}(t) + \int g_2(t, \mathbf{y}(t), \gamma)\nu(dt, d\gamma), \\ dy_3(t) = a_3(t, \mathbf{y}(t))dt + b_3(t, \mathbf{y}(t))d\mathrm{w}(t) + \int g_3(t, \mathbf{y}(t), \gamma)\nu(dt, d\gamma), \\ \mathbf{y}(0) = \mathbf{y}_0. \end{cases} \tag{30}$$

Further, we determine complementary function which is unit of control functions for PCP1:

$$\begin{cases} s_1(t, \mathbf{y}(t)) = a_1(t, \mathbf{y}(t)) - F_1(t, \mathbf{y}(t)), \\ s_2(t, \mathbf{y}(t)) = a_2(t, \mathbf{y}(t)) - F_2(t, \mathbf{y}(t)), \\ s_3(t, \mathbf{y}(t)) = a_3(t, \mathbf{y}(t)) - F_3(t, \mathbf{y}(t)). \end{cases} \tag{31}$$

Finally, we have constructed stochastic analogue for classical model described by a differential equations system and having an invariant function.

9.3 The SIR (susceptible-infected-recovered) model

The SIR is a simple mathematical model of epidemic [24], which divides the (fixed) population of N individuals into three" compartments" which may vary as a function of time t.

$S(t)$ are those susceptible but not yet infected with the disease,

$I(t)$ is the number of infectious individuals,

$R(t)$ are those individuals who have recovered from the disease and now have immunity to it,

the parameter λ describes the effective contact rate of the disease,

the parameter μ is the mean recovery rate.

The SIR model describes the change in the population of each of these compartments in terms of two parameters:

$$\begin{cases} \dfrac{dS(t)}{dt} = -\lambda \dfrac{S(t)I(t)}{N}, \\ \dfrac{dI(t)}{dt} = \lambda \dfrac{S(t)I(t)}{N} - \mu I(t), \\ \dfrac{dR(t)}{dt} = \mu I(t), \end{cases} \tag{32}$$

and its restrictsion is

$$S(t) + I(t) + R(t) = N. \tag{33}$$

Let the model with strong perturbation be

$$\begin{cases} dS(t) = \left(-\lambda \dfrac{S(t)I(t)}{N} + s_1(t, S(t), I(t), R(t))\right)dt + \\ \qquad +b_1(t, S(t), I(t), R(t))d\mathrm{w}(t) + \int_{R_\gamma} g_1(t, S(t), I(t), R(t), \gamma)\nu(dt, d\gamma), \\ dI(t) = \left(\lambda \dfrac{S(t)I(t)}{N} - \mu I(t) + s_2(t, S(t), I(t), R(t))\right)dt + \\ \qquad +b_1(t, S(t), I(t), R(t))d\mathrm{w}(t) + \int_{R_\gamma} g_2(t, S(t), I(t), R(t), \gamma)\nu(dt, d\gamma), \\ dR(t) = (\mu S(t) + s_3(t, S(t), I(t), R(t)))dt + \\ \qquad +b_1(t, S(t), I(t), R(t))d\mathrm{w}(t) + \int_{R_\gamma} g_3(t, S(t), I(t), R(t), \gamma)\nu(dt, d\gamma), \end{cases} \tag{34}$$

and

$$u(t, S(t), I(t), R(t)) = S(t) + I(t) + R(t) - N \equiv 0. \tag{35}$$

Suppose that the function $u(t, x, y, z) = x + y + z - N$ is a first integral, $v(t, x, y, z) = 2e^{-t} + x$ and $h(t, x, y, z) = y$ are complementary functions, and $q(t, x, y, z) = x$ is arbitrary function. The initial condition is: $x(0) = 1, y(0) = 0$, $z(0) = 0$. Then constructed differential equations system has the form

$$\begin{bmatrix} dx(t) \\ dy(t) \\ dz(t) \end{bmatrix} = \begin{bmatrix} 2e^{-t} \\ 0 \\ -2e^{-t} \end{bmatrix} dt + \begin{bmatrix} 0 \\ x(t) \\ -x(t) \end{bmatrix} d\mathrm{w}(t) + \int_{R_\gamma} \begin{bmatrix} 0 \\ x(t)\gamma) \\ -x(t)\gamma \end{bmatrix} \nu(dt, d\gamma). \tag{36}$$

Let us simulate a numerical solution of Eg.(36), where $N = 1$ (for example). **Figure 1** shows simulation for system without jumps, the **Figure 2** shows the processes with jumps.

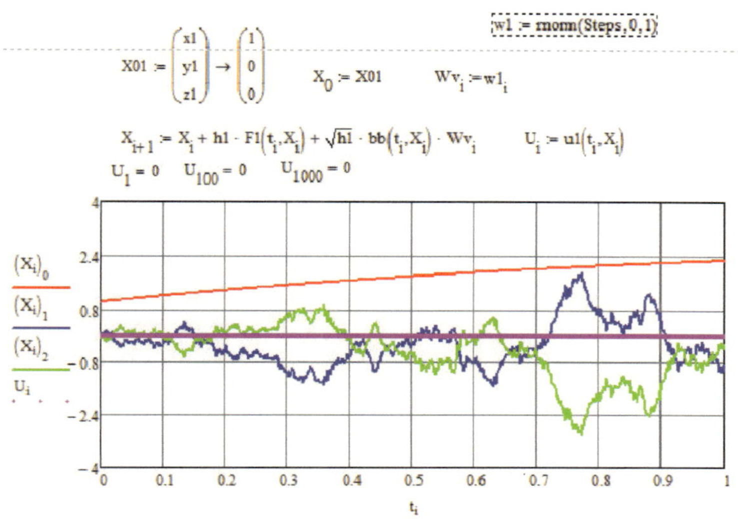

Figure 1.
Numerical solution for Eq.(36) without jumps.

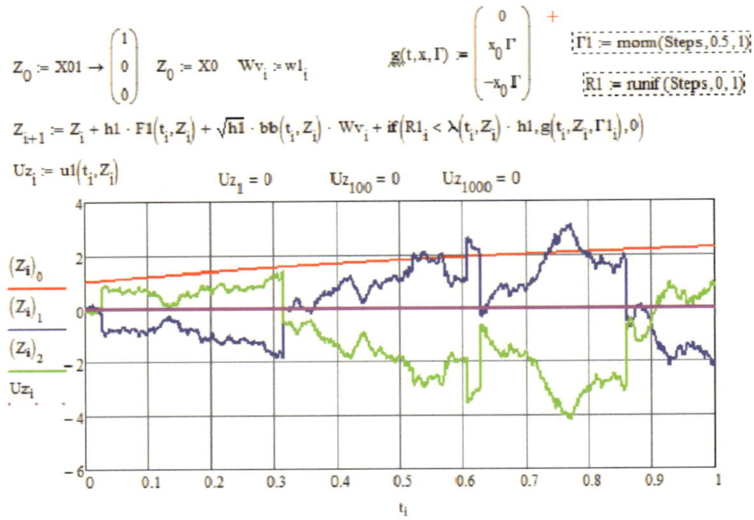

Figure 2.
Numerical solution for Eq.(36) with jumps.

In such a way we could use the system of differential equations

$$\begin{cases} dy_1(t) = 2e^{-t}dt, \\ dy_2(t) = y_1(t)dw(t) + \int \gamma y_1(t)\nu(dt, d\gamma), \\ dy_3(t) = -2e^{-t}dt + y_1(t)dw(t) - \int \gamma y_1(t)\nu(dt, d\gamma), \\ y(0) = y_0, \end{cases} \tag{37}$$

as initial step for construction of stochastic SIR-model. A good choice of complementary functions $v(t,x,y,z)$ and $h(t,x,y,z)$ allows us to obtain such coefficients that ensure that the solution $\{x(t), y(t), z(t)\}$ of the differential equations system satisfy some reasonable limitations.

9.4 The predator–prey model

The Lotka - Volterra equations or the predator–prey equations used to describe the dynamics of biological systems in which two species interact, one as a predator and the other as prey.

The Lotka - Volterra model makes a number of assumptions, not necessarily realizable in nature, about the environment and evolution of the predator and prey populations:

- The prey population finds ample food at all times.
- The food supply of the predator population depends entirely on the size of the prey population.
- The rate of change of population is proportional to its size.

- During the process, the environment does not change in favor of one species, and genetic adaptation is inconsequential.

- Predators have limitless appetite.

Let us note: $N_1(t)$ is the number of prey, and $N_2(t)$ is the number of some predator, ε_1, ε_2, η_1 and η_2 are positive real parameters describing the interaction of the two species.

The populations change through time according to the pair of equations:

$$\begin{cases} dN_1(t) = N_1(t)(\varepsilon_1 - \eta_1 N_2(t))dt, \\ dN_2(t) = -N_2(t)(\varepsilon_2 - \eta_2 N_1(t))dt. \end{cases} \tag{38}$$

Eq. (38) has the invariant function

$$N_1^{-\varepsilon_2}(t)e^{\eta_2 N_1(t)} = CN_2^{\varepsilon_1}(t)e^{-\eta_1 N_2(t)}, \tag{39}$$

where $C = const.$

We can introduce the stochastic model as a form

$$\begin{cases} dx_1(t) = (\varepsilon_1 x_1(t) - \eta_1 x_1(t)x_2(t) + s_1(t, x_1(t), x_2(t)))dt + \\ \qquad + b_1(t, x_1(t), x_2(t))dw_1(t) + \int_{R_\gamma} g_1(t, x_1(t), x_2(t), \gamma)\nu(dt, d\gamma), \\ dx_2(t) = (-\varepsilon_2 x_2(t) + \eta_2 x_1(t)x_2(t) + s_2(t, x_1(t), x_2(t)))dt + \\ \qquad + b_2(t, x_1(t), x_2(t))dw_2(t) + \int_{R_\gamma} g_2(t, x_1(t), x_2(t), \gamma)\nu(dt, d\gamma), \\ x_1(0) = N_1, \qquad x_2(0) = N_2, \end{cases} \tag{40}$$

with condition

$$u(t, \mathbf{x}(t)) = x_1^{-\varepsilon_2}(t)e^{\eta_2 x_1(t)} - Cx_2^{\varepsilon_1}(t)e^{-\eta_1 x_2(t)}. \tag{41}$$

Let us assume that $\varepsilon_1 = 2$, $\varepsilon_2 = 1$, $\eta_1 = \eta_2 = 1$, and $C = 1$, and initial condition is $x(0) = y(0) = 1$. The function $u(t, x, y) = x^{-1}e^x - y^2 e^{-2y}$ is a first integral, $h(t, x, y) = y - x + e^{-t}$ and $q(t, x, y) = x$ are complementary functions.

We cannot find an analytical solution of the differential equations system

$$\begin{cases} \dfrac{\partial z_1(t, x, y, \gamma)}{\partial \gamma} = e^{-z_2(t,x,y,\gamma)} z_1(t, x, y, \gamma) z_2(t, x, y, \gamma)(z_2(t, x, y, \gamma) - 2), \\ \dfrac{\partial z_2(t, x, y, \gamma)}{\partial \gamma} = -e^{-z_1(t,x,y,\gamma)}\left(1 - z_1^{-1}(t, x, y, \gamma)\right). \end{cases}$$

Then, the constructed SDE system includes only Wiener perturbation:

$$\begin{bmatrix} dx(t) \\ dy(t) \end{bmatrix} = \begin{bmatrix} A(t, x(t), y(t)) + B(t, x(t), y(t)) + C(t, x(t), y(t)) \\ D(t, x(t), y(t)) + E(t, x(t), y(t)) \end{bmatrix} dt + \\ + \begin{bmatrix} x(t)e^{-y(t)}(y^2(t) - 2y(t)) \\ \dfrac{e^{x(t)}}{x(t)} - e^{x(t)} \end{bmatrix} dw(t), \tag{42}$$

where

$$
\begin{aligned}
A(t,x(t),y(t)) &= 0.5e^{-y(t)}x(t)\left(y^2(t) - 2y(t)e^{-y(t)}\right)^2,\\
B(t,x(t),y(t)) &= 0.5e^{-y(t)}\left(x(t)e^{x(t)} - e^{x(t)}x^{-1}(t)\right)\left(2 - 4y(t) + y^2(t)\right),\\
C(t,x(t),y(t)) &= -\frac{e^{-t}e^{-y(t)}\left(y^2(t) - 2y(t)\right)}{e^{x(t)}x^{-1}(t) - 2y(t)e^{-y(t)} - e^{x(t)}x^{-2}(t) + y^2(t)e^{-y(t)}},\\
D(t,x(t),y(t)) &= \frac{e^{-t}e^{x(t)}\left(x^{-1}(t) - x^{-2}(t)\right)}{e^{x(t)}x^{-1}(t) - 2y(t)e^{-y(t)} - e^{x(t)}x^{-2}(t) + y^2(t)e^{-y(t)}},\\
E(t,x(t),y(t)) &= -0.5e^{x(t)}e^{-y(t)}\left(y^2(t) - 2y(t)\right)\left(1 - x^{-1}(t) + x(t) - 2 + 2x^{-1}(t)\right).
\end{aligned}
\tag{43}
$$

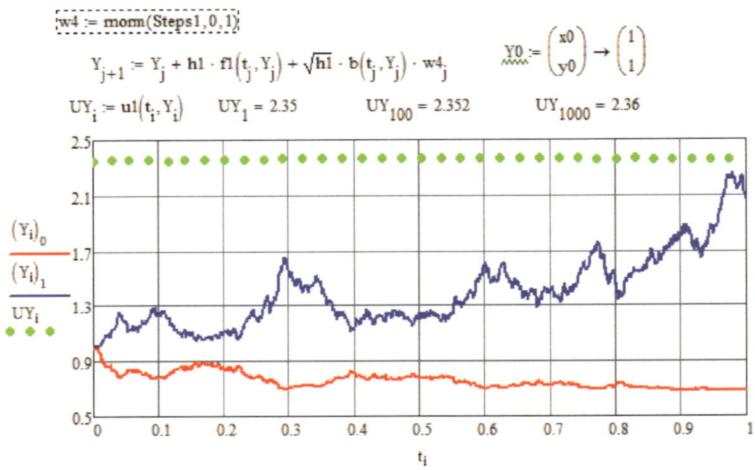

Figure 3.
Numerical simulation 1 for solution of Eq.(44).

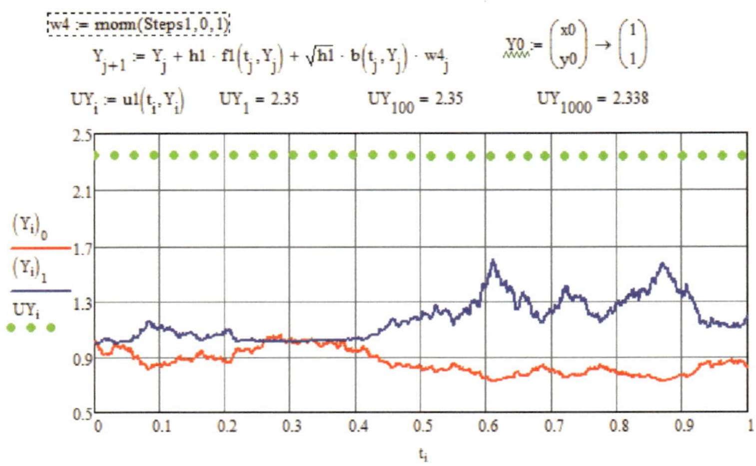

Figure 4.
Numerical simulation 2 for solution of Eq.(44).

Finally, we have the stochastic Lotka Volterra model associated to (38) $(N(t) = (N_1(t), N_2(t)))$:

$$\begin{bmatrix} dN_1(t) \\ dN_2(t) \end{bmatrix} = \begin{bmatrix} A(t, N(t)) + B(t, N(t)) + C(t, N(t)) \\ D(t, N(t)) + E(t, N(t)) \end{bmatrix} dt + \\ + \begin{bmatrix} N_1(t) e^{-N_2(t)} \left(N_2^2(t) - 2N_2(t)\right) \\ \dfrac{e^{N_1(t)}}{N_1(t)} - e^{N_1(t)} \end{bmatrix} dw(t), \tag{44}$$

where $A(t, N(t))$, $B(t, N(t))$, $C(t, N(t))$, $D(t, N(t))$, $E(t, N(t))$ are determined by Eq.(43).

Figures 3 and **4** show two realizations for numerical solution of Eq. (44).

Another examples of a differential equation system construction and models see in [25–29].

10. Conclusion

The invariant method widens horizons for constructing and researching into mathematical models of real systems with the invariants that hold out under any strong random disturbances.

Author details

Elena Karachanskaya
Far-Eastern State Transport University, Khabarovsk, Russia

*Address all correspondence to: elena_chal@mail.ru

References

[1] Dubko V. A. First Integrals of Systems of Stochastic Differenial Equations. Preprint. 1978. Izd-vo An USSR, In-t Matematiki, Kiev. [in Russian], https://ru.calameo.com/read/003168372f1f76b3dbbbc [Accessed: 08 January 2021]

[2] Dubko V. A. Integral invariants for some class of systems of stochastic differential equations. Dokl. Akad. Nauk Ukr. SSR, Ser. A. 1984, **18**(1) : 17–20. [In Ukranian]

[3] Dubko V. A. Questions in the Theory and Applications of Stochastic Differential Equations. DVO AN SSSR, Vladivostok, 1989. [in Russian].

[4] Dubko V. A. and Chalykh E. V. Construction of an analytic solution for one class of equations of the Langevin type with orthogonal random actions. Ukr. Mat. Zh. 1998, **50**(4) : 666–668. https://doi.org/10.1007/BF02487397

[5] Chalykh E. On one generalization of the Langevin equation with determinate modulus of velocity. Ukr. Mat. Zh. 1998, **50**(7) : 1004–1006. https://doi.org/10.1007/BF02528827

[6] Dubko V. A. Integral invariants of Itôs equations and their relation with some problems of the theory of stochastic processes. Dopov. Nats. Akad. Nauk Ukr.,Mat. Pryr. Tekh. Nauky. 2002, 1 : 24–27.

[7] Dubko V. A. Open evolving systems. In: The First International Scientific Technical Conference Open Evolving Systems (April 2627, 2002), Abstracts, 14 (VNZ VMURoL, Kyiv, 2002), http://openevolvingsystems.narod.ru/indexUkr.htm [Accessed: 08 January 2021]

[8] Doobko V. A. and Karachanskaya E. V. Stochastic first integrals, kernel of integral invariants and Kolmogorov equations. Far Eastern Mathematical Journal. 2014, , **14**(2) : 1–17. [In Russian] https://readera.org/14329339-en [Accessed: 08 January 2021]

[9] Karachanskaya E.V. The generalized ItôVenttsel formula in the case of a noncentered Poisson measure, a stochastic first integral, and a first integral. Sib. Adv. Math. 2015, **25** : 191–205. https://doi.org/10.3103/S1055134415030049 [Accessed: 08 January 2021]

[10] Karachanskaya E. V. Random processes with invariant. Khabarovsk, Publ of Nat. Pacific Univ. 2014, P. 148. [In Russian] http://lib.pnu.edu.ru/downloads/TextExt/uchposob/Karachanskaya.pdf?id=1156454 [Accessed: 08 January 2021]

[11] Erugin N. P. Construction of the whole set of differential equations having a given integral curve. Prikl. Mat. Mekh. 1952, **16**(6):658–670. [In Russian]

[12] Gikhman I. I. and Skorokhod A. V. Stochastic Differential Equations. Springer-Verlag, New York, 1972: viii + 354 pp. (Translation of the 1968 book, Stokhasticheskie Differentsialnie Uravneniya. Published by Naukova Dumka, Kiev, U.S.S.R.)

[13] Itô K. Stochastic differential equations in a differentiable manifold. Nagoya Math. J. 1950, **1** : 35–47.

[14] Kunita H. and Watanabe S. On square integrable martingales. Nagoya Math. J. 1967, **30** : 209–245. https://doi.org/10.1017/S0027763000012484 [Accessed: 08 January 2021]

[15] Ventzel' A. D. On equations of the theory of conditional Markov processes. Theory Probab. Appl. 1965, **10**(2) : 357–361. Translated in Russian : Teor. Veroyatnost. i Primenen. 1965, **10**(2) : 390–393. https://doi.org/10.1137/1110045

[16] Karachanskaya E. V. A Proof of the Generalized Itô – Wentzell Formula via the Delta-Function and the Density of

Normal Distribution. Yakutian Math. J. 2014, **21**(3): 46–59 http://mzsvfu.ru/index.php/mz/article/view/a-proof-of-the-generalized-ito-wentzell-formula/194 [Accessed: 08 January 2021]

[17] Karachanskaya E.V. A direct method to prove the generalized Itô - Venttsel' formula for a generalized stochastic differential equation, Siberian Adv. Math. 2016, **26**(1): 17–29. https://doi.org/10.3103/S1055134416010028

[18] Karachanskaya E. V. Construction of program controls with probability 1 for a dynamical system with Poisson perturbations. Vestnik Tikhookeansk. Gosuniversiteta 2011, **2**(21): 51–60. [In Russian]. https://pnu.edu.ru/media/vestnik/articles/1586/en/ [Accessed: 08 January 2021]

[19] Karachanskaya E. V. Programmed Control with Probability 1 for Stochastic Dynamical Systems. Journal of Mathematical Sciences. 2020, **248**(2) : 67–79. https://link.springer.com/article/10.1007/s10958-020-04856-4 [Accessed: 08 January 2021]

[20] Krylov N. V. and Rozovskii B. L. Stochastic partial differential equations and diffusion processes. Usp. Mat. Nauk. 1982, **37**(6 (228)) : 75–87 [Russ. Math. Surv. 1982, **37**: 81–89].

[21] Poincare A. Selected Works in Three Volumes. Vol. II: Les Methodes Nouvelles de la Mecanique Celeste. Topology. Number Theory. Nauka, Moscow. 1972.

[22] Karachanskaya E. V. Construction of Programmed Controls for a Dynamic System Based on the Set of its First Integrals. Journal of Mathematical Sciences 2014, **199**(5) : 547–555. https://doi.org/10.1007/s10958-014-1881-4 [Accessed: 08 January 2021]

[23] Karachanskaya E. V. Stochastization of classical models with dynamical invariants. Mathematical notes of NEFU. 2020, **27**(1): 69–87. [In Russin] http://mzsvfu.ru/index.php/mz/article/view/stochastization-of-classical-models-with-dynamical-invariants [Accessed: 08 January 2021]

[24] Kermack W. O. and McKendrick A. G. Contributions to the mathematical theory of epidemics. Proc. R. Soc. Lond., Ser. A, Math. Phys. Eng. Sci. 1927, **115** (772) : 700–721.

[25] Chalykh E. V. Constructing the set of program controls with probability 1 for one class of stochastic systems. Automation and Remote Control. 2009, **70**(8) : 1364–1375. https://link.springer.com/article/10.1134/S0005117909080098 [Accessed: 08 January 2021]

[26] Karachanskaya E. and Tagirova T. Construction of stochastic transport models with a constant function. IOP Conference Series Earth and Environmental Science. December 2019, **403**, no. 012211. https://iopscience.iop.org/article/10.1088/1755-1315/403/1/012211 [Accessed: 08 January 2021]

[27] Karachanskaya E. V. and Petrova A. P. Modeling of the programmed control with probability 1 for some financial tasks. Mathematical notes of NEFU. 2018, **25**(1): 25–37. [In Russin] https://doi.org/10.25587/SVFU.2018.1.12766 [Accessed: 08 January 2021]

[28] Averina T., Karachanskaya E., and Rybakov K. Statistical modeling of random processes with invariants. International Multi-Conference on Engineering, Computer and Information Sciences (SIBIRCON). Novosibirsk. 2017: 34–37. https://ieeexplore.ieee.org/document/8109832 [Accessed: 08 January 2021]

[29] Averina T. A., Karachanskaya E. V., and Rybakov K. A. Statistical analysis of diffusion systems with invariants. Russian Journal of Numerical Analysis and Math. Modelling. 2018, **33**(1) : 1–13. https://doi.org/10.1515/rnam-2018-0001 [Accessed: 08 January 2021]

Chapter 5

Stochastic Theory of Coarse-Grained Deterministic Systems: Martingales and Markov Approximations

Michel Moreau and Bernard Gaveau

Abstract

Many works have been devoted to show that Thermodynamics and Statistical Physics can be rigorously deduced from an exact underlying classical Hamiltonian dynamics, and to resolve the related paradoxes. In particular, the concept of equilibrium state and the derivation of Master Equations should result from purely Hamiltonian considerations. In this chapter, we reexamine this problem, following the point of view developed by Kolmogorov more than 60 years ago, in great part known from the work published by Arnold and Avez in 1967. Our setting is a discrete time dynamical system, namely the successive iterations of a measure-preserving mapping on a measure space, generalizing Hamiltonian dynamics in phase space. Using the notion of Kolmogorov entropy and martingale theory, we prove that a coarse-grained description both in space and in time leads to an approximate Master Equation satisfied by the probability distribution of partial histories of the coarse-grained state.

Keywords: stochastic theory, coarse-grained deterministic systems, Markov processes, martingales

1. Introduction

It is generally admitted that Thermodynamics and Statistical Physics could be deduced from an exact classical or quantum Hamiltonian dynamics, so that the various paradoxes related to irreversibility could also be explained, and nonequilibrium situations could be rigorously studied as well. These questions have been and still are discussed by many authors (see, for instance Refs. [1–4] and many classical textbooks, for instance [5–10]), who have introduced various plausible hypotheses [7–14], related to the ergodic principle [8–11], to solve them. It seems that there are two major kinds of problems. First, to justify that physical systems can reach an equilibrium state when they are isolated, or in contact with a thermal bath (which remains to be defined). Secondly, to justify various types of reduced stochastic dynamics, depending on the phenomena to be described: Boltzmann equations, Brownian motions, fluid dynamics, Bogoliubov-Born-Green-Kirkwood-Yvon (BBGKY) hierarchies, etc..: see for instance Refs [1–11, 15, 16]. Concerning the first type of problems (reaching an equilibrium, if any) very rough estimations show [17] that the time scales to reach equilibrium, using only Hamiltonian dynamics and measure inaccuracies, are extremely large, contrarily to everyday experience, and quantum estimations are even worse [17]. Essentially, these times scale as Poincaré recurrence times and they increase as an exponential of the number of degrees of freedom (see Section 3 of this chapter for a brief discussion and references).

Here we concentrate on the second type of problems: is it possible to derive a stochastic Markovian process from an "exact" deterministic dynamics, just by coarse graining the microscopic state space? We generalize and complete the formalism recently presented [18] for Hamiltonian systems. Our framework is now more general and applies to all deterministic systems with a measure preserving dynamics, which, by Liouville theorem, include Hamiltonian dynamics.

Following Kolmogorov, we start with a measure space with a discrete time dynamics given by the successive iterations of a measure preserving mapping. The Kolmogorov entropy, or trajectory entropy, has been defined by Kolmogorov as an invariant of stationary dynamical systems (see Arnold and Avez book [19] for a pedagogical presentation). We follow his work and generalize part of his results. We also use martingale theory [20–23] to show that the stationary coarse-grained process almost surely tends to a Markov process on partial histories including n successive times, when n tends to infinity. From this result, we show that in the nonstationary situation, the probability distribution of such partial histories approximately satisfies a Master equation. Its probability transitions can be computed from the stationary distribution, expressed in terms of the invariant measure. It follows that, with relevant hypotheses, the mesoscopic distribution indeed tends to the stationary distribution, as expected.

Our next step is to coarse grain time also. The new, coarse-grained time step is now $n\,\tau$, τ being the elementary time step of the microscopic description, and n being the number of elementary steps necessary to approximately "erase" the memory with a given accuracy. The microscopic dynamics induces new dynamics on partial histories of length n. We show that it is approximately Markovian if n is large enough. This idea is a generalization of the Brownian concept: a particle in a fluid is submitted to a white noise force which is the result of the coarse-graining of many collisions, and the time step is thus the coarse-graining of many microscopic time steps [8, 24]. The Brownian motion emerges as a time coarse-grained dynamics.

In Section 2, we recall various mathematical concepts (Kolmogorov entropy, martingale theory) and use them to derive the approximate Markov property of the partial histories, and eventually to obtain an approximate Master Equation for the time coarse-grained mesoscopic distribution [18].

In Section 3, we briefly consider the problem of relaxation times and recall very rough estimations showing that an exact Hamiltonian dynamics predicts unrealistic, excessively large relaxation times [17], unless the description is completed by introducing other sources of randomness than the measure inaccuracies leading to space coarse-graining. Note that, following Kolmogorov [19], we do not address the Quantum Mechanics formalism.

2. Microscopic and mesoscopic processes in deterministic dynamics

2.1 Microscopic dynamics: Definitions and notations

It has been shown recently [18] that coarse-grained Hamiltonian systems can be approximated by Markov processes provided that they satisfy reasonable properties, covering many realistic cases. These conclusions can be extended to a large class of deterministic systems generalizing classical Hamiltonian systems, which we now describe. We first specify our hypotheses and notations.

2.1.1 Deterministic microdynamics

Consider a deterministic system $\mathcal{S}$. Its states x, belonging to a state space X, will be called "microstates", in agreement with the usual vocabulary of Statistical Physics. The deterministic trajectory due to the microscopic dynamics transfers the microstate x_0 at time 0 to the microstate $x_t = \varphi_t(x_0)$ at time t. The evolution function φ_t satisfies the current properties of dynamic systems: $\varphi_t\, \varphi_s = \varphi_{t+s}, \varphi_0 = I$, t and s being real numbers and I being the identical function.

The dynamics is often invariant by time reversion, as assumed in many works on Statistical Physics: we refer to classical textbooks on the subject for details [5–8], but we will not use such properties in this chapter.

2.1.2 Microscopic distribution

Assume that the exact microstate x_0 is unknown at time 0, but is distributed according to the probability measure μ on the phase space X. The microscopic probability distribution μ_t at time t is given by

$$\mu_t\,(A) = \mu(\varphi_{-t}\, A). \tag{1}$$

for any measurable subset A of X. If μ is stationary, it is preserved by the dynamics: $\mu_t\,(A) = \mu\,(A)$. This condition, however, it not necessarily satisfied, in particular for physical systems during their evolution.

We will focus on two important cases:

a. the finite case: X is finite and consists in N microstates.

b. the absolutely continuous case: $X \subset R^n$, where *(i)* R is the set of real numbers and n is an integer (usually very large, and even in the case of Hamiltonian dynamics), and *(ii)* the measure μ is absolutely continuous with respect to the Lebesgue measure ω on R^n: these exists an integrable probability density $p(x)$ such that for any measurable subset A of X

$$\mu(A) = \int_A p(x) d\omega(x). \tag{2}$$

Furthermore, we assume that *(iii)* the Lebesgue measure of X (or volume of X) $V = \text{vol}(X) \equiv \int_X d\omega(x)$ is finite, and *(iv)* the Lebesgue measure ω is preserved by the dynamics for any t and any measurable subset A of X.:

$$\text{vol} A \;= \text{vol}\;(\varphi_{-t} A). \tag{3}$$

The last two assumptions obviously generalize basic properties of Hamiltonian dynamics in a finite volume of phase space. Thus, by (1)–(3), the probability density is conserved along any trajectory: at time t the probability density is

$$p(x,t) \;= p\,(0, \varphi_{-t} x) \equiv p\,(\varphi_{-t} x). \tag{4}$$

2.1.3 Initial microscopic distribution: The stationary situation

Suppose that $\mathcal{S}$ is an isolated physical system and no observation was made on $\mathcal{S}$ at time 0 nor before 0. Then, in the absence of any knowledge on $\mathcal{S}$, we admit that at the initial time $\mathcal{S}$ is distributed according to the only unbiased probability law, which is the uniform law. This is clearly justified in the finite case, according to the physical meaning traditionally given to probability: in fact, attributing different probabilities for two distinct microstates of X would imply that some measurement would allow one to distinguish them objectively, which is not the case at time 0.

In the absolutely continuous case, initial uniformity is less obvious: it amounts to assuming that the system should be found with equal probability in two regions of the state space with equal volumes if no information allows one to give preference to any of these regions. This is of course a subjective assertion, but for Hamiltonian systems it agrees with the semi-quantum principle which asserts that, in canonical coordinates, equal volumes of the phase space correspond to equal numbers of quantum states.

Another way for choosing the initial probability distribution is to make use of Jaynes' principle [25], which is to maximize the Shannon entropy of the distribution under the known constraints over this distribution: in the present case of an isolated system which has not been previously observed, this principle also leads to the uniform law. It is not really better founded than the previous, elementary reasoning, but it may be more satisfying and it can be safely used in more complex situations. We refer to most textbooks on statistical mechanics for discussing these well-known, basic questions.

The uniform distribution in a finite space, either discrete or absolutely continuous, is clearly stationary. In addition to the previous hypotheses, we will assume that the space X is indecomposable [26]: the only subsets of X which are preserved by the evolution function φ_t are the empty set $\varnothing$ and X itself. Then, the stationary probability distribution is unique [18].

For simplicity, we will henceforth assume that the phase space X is finite.

Initial, nonstationary situation. In certain situations, the system can be prepared by submitting it to specific constraints before the initial time 0. Then it may not be distributed uniformly in X at $t = 0$. We will consider this case in the next paragraph.

2.2 Mesoscopic distributions

2.2.1 Mesoscopic states

Because of the imprecision of the physical observations, it is impossible to determine exactly the microstate of the system, but it is currently admitted that the available measure instruments allow one to define a finite partition of X into subsets $i \in \mathcal{M} \equiv (i^k)$, $k = 1, 2, \ldots M$, such that it is impossible to distinguish two microstates belonging to the same subset i. So, in practice the best possible description of the system consists in specifying the subset i where its microstate x lies: i can be called the *mesostate* of the system. The probability for the system to be in the mesostate i at time t will be denoted $p(i,t)$. It is not sure, however, that two microstates belonging two different mesostates can always be distinguished: this point will be considered in Section 3.2.2.

Remark: for convenience, we use the same letter p to denote the probability in a countable state space, as well as the probability density in the continuous case. This creates no confusion when the variable type is explicitly mentioned. This is the case now since, as mentioned previously, we assume that the space X is discrete. The transposition to the continuous case is generally obvious, although the complete derivations may be more difficult.

2.2.2 The stationary situation

If time 0 is the beginning of all observations and actions, we assume that the initial microscopic distribution μ is uniform and stationary, as discussed previously, and the probability to find system $\mathcal{S}$ in the mesostate i_0 at time 0 is $p(i_0, 0) = \mu(i_0,)$. The probability to be in i at time t is $p^0(i,t) = \mu_t(i) = \mu(\varphi_{-t}(i))$. The stationary joint probability to find $\mathcal{S}$ in i_0 at time 0 and in i at time t is

$$p^0(i_0, 0; i, t)) = \mu(\varphi_{-t} i \cap i_0) = \mu(i \cap \varphi_t i_0) \tag{5}$$

and the conditional probability of finding $\mathcal{S}$ in the i at time t, knowing that it was in i_0 at time 0 is

$$p^0(i, t|i_0, 0) = \frac{p^0(i, t; i_0, 0)}{p^0(i_0, 0)} = \frac{\mu(\varphi_{-t}\, i \cap i_0)}{\mu(i_0)} = \frac{\mu(\, i \cap \varphi_t i_0)}{\mu(i_0)} \tag{6}$$

Similarly, the stationary n-times joint probability and related conditional probabilities are readily obtained from

$$p^0(i_0, 0; i_1, t_1; \ldots i_{n-1}, t_{n-1}) \;=\; \mu\left(\varphi_{-t_{n-1}} i_{t_{n-1}} \cap \ldots \cap i_0\right) . \tag{7}$$

with, for any t: $p^0(i_0, t; i_1, t_1 + t; \ldots i_{n-1}, t_{n-1} + t) = p^0(i_0, 0; i_1, t_1; \ldots\, i_{n-1}, t_{n-1})$.

For the sake of simplicity, we will discretize the times $0 < t_1 < t_2 \ldots$, and write $t_i = k_i \tau$, k_i being a nonnegative integer and τ a constant time step, which will be taken as time unit.

2.2.3 Non stationary situation

If $\mathcal{S}$ is a physical system, interactions may exist before or at time 0, so that the $\mathcal{S}$ can be constrained to lie in a certain subset A of X at time 0. However, since it is not possible to distinguish two microstates corresponding to the same mesostate, A should be a union of mesostates, or at least one mesostate. If it is known that at time 0 the microsate x of the system belongs to the mesostate i, we should assume that the initial microscopic distribution is uniform over i, since no available observation can give further information on x: so, in the discrete case, if $n(i)$ is the number of microstates included in i and χ_i (x) the characteristic function of i

$$p(x, 0|x \in i) = \frac{1}{n(i)}\, \chi_i(x) \tag{8}$$

In the absolutely continuous case, the similar conditional density is obtained in the same way, replacing the number of microscopic states contained in the mesostate i by its volume $v(i)$. For simplicity, we follow considering the discrete case, with obvious adaptations to the continuous case.

If one only knows the mesoscopic initial distribution $p(i,0)$ that at time 0 the system belongs to i, for each mesostate i of $\mathcal{M}$, the initial microscopic distribution becomes

$$p(x, 0) \;= \sum\nolimits_i \frac{1}{n(i)} p(i, 0)\, \chi_i(x) = \sum\nolimits_i \frac{p(i, 0)}{\mu(i)} \frac{\chi_i(x)}{N} \tag{9}$$

N being the total number of microstates in X.

The n-times nonstationary mesoscopic probabilities are obtained from (9)

$$p_n(i_0, 0; i_1, 1, ; i_{n-1}, n-1) = p(i_0 \cap \varphi_{-1} i_1 \cap \ldots \cap \varphi_{-n+1} i_{n-1}, 0) = \frac{p(i_0)}{\mu(i_0)} \frac{n(i_0 \cap \varphi_{-1} i_1 \cap \ldots \cap \varphi_{-n+1} i_{n-1})}{N}. \quad (10)$$

where $n(A)$ is the number of microstates belonging to some subset A of X. So

$$p_n(i_0, 0; i_1, 1, ; i_{n-1}, n-1) = \mu(i_0 \cap \varphi_{-1} i_1 \cap \ldots \cap \varphi_{-n+1} i_{n-1}) \frac{p(i_0)}{\mu(i_0)}. \quad (11)$$

and all multiple probabilities follow, for instance

$$p_{n-1}(i_1, 1; \ldots; i_n, n) = \sum_{i_0} \mu(i_0 \cap \varphi_{-1} i_1 \cap \ldots \cap \varphi_{-n} i_n) \frac{p(i_0)}{\mu(i_0)}. \quad (12)$$

The corresponding process is generally not Markovian. For instance, if $i_0 \cap \phi_{-1} i_1 \neq \varnothing$, $i_1 \cap \phi_{-1} i_2 \neq \varnothing$ and $i_0 \cap \phi_{-2} i_2 = \varnothing$, it is easily seen that $p(.i_2, 2|i_1, 1; i_0, 0) = 0$ but $p(i_2, 2|i_1, 1) \neq 0$.

From the definition of the relative probabilities, one can formally write

$$p(i_2, t_2) = \sum_{i_1} p(i_2, t_2 | i_1, t_1) p(i_1, t_1). \quad (13)$$

but in general this equation is useless, since the conditional probability $p(i2, t2| i1, t1)$ cannot be computed independently of $p(i1, t1)$.

It results from (11) that the nonstationary conditional probabilities, *conditioned by the whole past up to time* 0, are identical to the corresponding stationary probabilities: as an example

$$p(i_n, n | i_{n-1}, n-1; \ldots; i_0, 0) = p^0(i_n, n | i_{n-1}, n-1; \ldots; i_0, 0) = \frac{\mu(i_0 \cap \varphi_{-1} i_1 \cap \ldots \cap \varphi_{-n} i_n)}{\mu(i_0 \cap \varphi_{-1} i_1 \cap \ldots \cap \varphi_{-n+1} i_{n-1})}. \quad (14)$$

We will make use of this simple but important property later.

2.3 Entropy of the mesoscopic process resulting from deterministic, microscopic system

Kolmogorov and other authors [19] studied the entropy and ergodic properties of the stationary mesoscopic process defined previously, following methods introduced by Shannon in the framework of signal theory [27–30]. These methods, and part of Kolmogorov's results, can be extended to the nonstationary process (11).

2.3.1 The n-times entropy and the instantaneous entropy of the mesoscopic system

Following Kolmogorov, we consider y the Shannon entropy [27–30] of the trajectory $(i)_n = (i_0, \ldots, i_{n-1})$ in the phase space

$$S(p_n) = - \sum_{i_0, \ldots i_{n-1}} p_n(i_0, 0; \ldots; i_{n-1}, n-1) \ln p_n(i_0, 0; \ldots; i_{n-1}, n-1). \quad (15)$$

On the other hand, the new information obtained by observing the system in the mesoscopic state i_n at time t_n, knowing that it was in the respective states $i_0, \ldots i_{n-1}$ at the prior times $0, \ldots n\text{-}1$, will be called the instantaneous entropy

$$\begin{aligned} s_n(p) = S_{n+1}(p) - S_n(p) &= -\sum_{i_0,;;;i_n} p(i_0, 0; \ldots . i_n, n) \ln p(i_n, n | i_{n-1}, n-1; \ldots; i_0, 0) \geq 0 \\ &= \sum_{i_0,;;;i_{n-1}} p(i_0, 0; \ldots . i_{n-1}, n-1)\, S(p(\bullet, n | i_{n-1}, n-1; \ldots; i_0, 0)). \end{aligned} \tag{16}$$

where p denotes the infinite process. The properties of $S(p_n)$ and $s_n(p)$ have been extensively studied by Kolmogorov and other authors in the case of the stationary process (6) [19]: they are summarily mentioned in 2.5. They are not necessarily valid for the nonstationary process.

2.3.2 Maximizing the n*-times entropy of the mesoscopic system: The "Markov scheme"*

If one knows the first two distributions p_1 and p_2, one can mimics the exact mesoscopic distributions p_n by using the Jaynes' principle, maximizing the entropy $S(q_n)$ of a ditribution q_n under the constraints $q_1 = p_1$ and $q_2 = p_2$. Then it is found that optimal distribution q_n is the Markov distribution $\overline{q}_n$ satifying these constraints [18].

It is shown in Ref. [18] that for $n > 2$, both the n-times entropy $S_n(\overline{q})$ and the instantaneous entropy $s_n(\overline{q})$ are larger than the correponding entropies $S_n(p)$ and $s_n(p)$ of the exact process p, except if p is Markov: $p = \overline{q}$.

The Markov process $\overline{q}_n$ is not really an approximation of the mesoscopic process p, because $\overline{q}_n$ does not tend to p_n when $n \to \infty$. Approximating the exact mesoscopic process by a Markov process will be the main purpose of the next section.

2.4 Entropy and memory in the stationary situation

2.4.1 Kolmogorov entropy of the stationary process

Here we consider the stationary process arising from the initial uniform microscopic distribution $\mu(x)$, when the n-times stationary probability is p_n^0 given by (7). For the sake of simplicity we omit the index ^{0} in the present Section, unless otherwise specified. It can be shown [19] that the entropy $S_n(p)$ is an increasing, concave function of n

$$s_n \equiv S_{n+1}(p) - S_n(p) \geq 0. \tag{17}$$

$$s_{n+1} - s_n = S_{n+1}(p) - 2S_n(p) + S_{n-1}(p) \leq 0. \tag{18}$$

It results from (17) and (18), and also from 2.5.2, that the limits

$$\lim_{n\to\infty} \frac{1}{n} S_n(p) = \lim_{n\to\infty} s_n(p) = s(p). \tag{19}$$

exist: $s(p)$ is the Kolmogorov entropy of the evolution function f with respect to the partition (i) of the mesoscopic states [19]. More simply, we can call it entropy of the mesoscopic process.

2.4.2 Memory decrease in the stationary mesoscopic process

It has been proved recently [18] that, although it is infinite, the memory of the mesoscopic process fades out with time: for n large enough, if $N > n$ the probability of i_N at time N conditioned by the n last events, is practically equal to the probability at time N, conditioned by the *whole past* down to time 0.

$$p(i_N, N|\, i_{N-1}, N-1; \ldots i_n, N-n) \approx p(i_N, N|\, i_{N-1}, N-1; \ldots i_0, 0) \text{ when } n \to \infty. \tag{20}$$

More precisely, for any $\varepsilon > 0$, there exists a positive integer n such that for any $N > n$

$$0 < s_N < s_n < \varepsilon. \tag{21}$$

where s_n is the instantaneous entropy given by (14). In fact, let us write

$$\Pi_N(i_N) = p(i_N, N|i_{N-1}, N-1; \ldots\ldots i_0, 0) = \mu\left(f_{-N} i_N \middle|\, f_{-N+1} i_{N-1} \cap \ldots \cap i_0\right); \tag{22}$$

$$\Pi_N^{(n)}(i_N) = p(i_N, N|i_{N-1}, N-1; \ldots\ldots; i_{N-n}, N-n) = \mu\left(f_{-N} i_N \middle|\, f_{-N+1} i_{N-1} \cap \ldots \cap f_{-N+n} i_{N-n}\right). \tag{23}$$

For a given n, formula (23) allows one to define a new process $p^{(n)}$ from the original process p, which can be called "the approximate process of order n" of p (see Section 2.6). It results from (21) and from the stationarity of p that for any $\varepsilon > 0$, there is an integer $n(\varepsilon)$ depending only on ε, such that for any integers N, $n > n(\varepsilon)$

$$0 < s_n(p) - s_N(p) = \sum_{i_0, \ldots i_{N-1}} \mu\left(i_0 \cap f_{-1}(i_1) \cap \ldots \cap f_{-N+1}(i_{N-1})\right) S_{0,\ldots N-1}\left(\Pi_N \middle| \Pi_N^{(n)}\right) < \varepsilon. \tag{24}$$

where $S_{0,\ldots N-1}\left(\Pi_N | \Pi_N^{(n)}\right)$ is the relative entropy of Π_N with respect to $\Pi_N^{(n)}$: the last right hand member of Eq. (22) is the average of this relative entropy on the past of N. Because $s_N(p)$ decreases to a limit $\bar{s}(p)$ when $N \to \infty$, it results that

$$0 < \delta s_n(p) \equiv s_n(p) - \bar{s}(p) \le \varepsilon \text{ if } n > n(\varepsilon). \tag{25}$$

The total variation distance $d(P,Q)$ between two distributions P_j and Q_j over the states j of a finite set (j) is

$$d(P, Q) = \frac{1}{2} \sum_j \left|P_j - Q_j\right|. \tag{26}$$

Then, the total variation distance $d_{0,\ldots N-1}\left(\Pi_N, \Pi_N^{(n)}\right)$ between Π_N and $\Pi_N^{(n)}$ (for a given past trajectory between times 0 and N-1) is related to the relative entropy [18, 31] and it can be concluded that

$$\left\langle d_{0,\ldots N-1}\left(\Pi_N, \Pi_N^{(n)}\right) \right\rangle^2 \le \left\langle \left[d_{0,\ldots N-1}\left(\Pi_N, \Pi_N^{(n)}\right)\right]^2 \right\rangle < \varepsilon/2 \text{ if } n(\varepsilon) < n < N. \tag{27}$$

2.4.3 Convergence properties of the approximate process

Let us write $m = N\text{-}n > 0$. It follows [18] from (25) that for any fixed m, the total variation distance between the exact and the approximate probabilities $d_{0,\ldots m+n-1}\left(\Pi_{m+n}, \Pi_{m+n}^{(n)}\right)$ *tends to* 0 *in probability* when $n \to \infty$

DOI: http://dx.doi.org/10.5772/intechopen.95903

$$d_{0,\dots m+n-1}\left(\Pi_{m+n},\Pi_{m+n}^{(n)}\right) \xrightarrow{p} 0 \text{ if } n \to \infty. \tag{28}$$

So, the probability that this distance exceeds a given accuracy $a > 0$ can be made as small as desired by choosing n large enough.

Further results can be obtained by newly using the sationnarity of process p. In fact, it can be shown [18] that p is a martingale [20–22]. Then, general results from martingales theory (see below) show that when $n \to \propto$ the distance between the stationary conditional probability Π_{m+n} and its approximation $\Pi_{m+n}^{(n)}$ tends to 0 *almost surely* [18], as well as and in probability

$$d_{0,\dots m+n-1}\left(\Pi_{m+n},\Pi_{m+n}^{(n)}\right) \xrightarrow{a.s.} 0 \text{ if } n \to \infty. \tag{29}$$

So, the approximation $\Pi_{m+n}^{(n)}$ converges to Π_{m+n} for almost all trajectories [18]. We now sketch the derivation of this conclusion from martingale theory.

2.5 Martingale theory and almost sure convergence

For convenience, we first summarize some definitions and results of martingale theory [20–22], before applying them to the mesoscopic laws of deterministic systems. We refer to [20] for adressing more general cases.

2.5.1 Definitions

i. *simplified definition*: a (discrete time) sequence of stochastic variables X_n is a martingale if for all n:

$$\langle |X_n| \rangle < \infty \text{ and } \langle X_{n+1} | X_n, \dots X_1 \rangle = X_n. \tag{30}$$

where $\langle X \rangle$ denotes the average (mathematical expectation) of the stochastic variable X.

ii. *more generally* (see the general definition, for instance, in [20])

If • $(\Omega, \mathcal{F}, P)$ is a probability space (where Ω is the state space, P is the probability law, and $\mathcal{F}$ is the set of all subspaces (σ-algebra) for which P is defined),

- $\mathcal{F}_n$ is an increasing sequence of σ-algebras extracted from $\mathcal{F}$ ($\mathcal{F}_n \subset \mathcal{F}_{n+1} \subset \dots \subset \mathcal{F}$), and.

- for all $n \geq 0$, X_n is a stochastic variable defined on $(\Omega, \mathcal{F}_n, P)$,

the sequence X_n is a martingale if $\langle |X_n| \rangle < \infty$ and $\langle X_{n+1} | \mathcal{F}_n \rangle = X_n$.

2.5.2 Convergence theorem for martingales

Among the remarkable properties of martingales, the following convergence theorem holds [20, 21]:

If (X_n) is a positive martingale, the sequence X_n converges almost surely to a stochastic variable X.

So, for almost all trajectories ω, $X_n(\omega) \to X(\omega)$ with probability 1 when $n \to \infty$. Stronger and more general results can be found in the references.

2.5.3 Application to the nth approximation of the stationary mesocopic process

The stochastic variable $Y_N = p^0(i_N, N|i_{N-1}, N-1; \ldots; i_0, 0)$ is a martingale. In fact, because of the stationarity of p^0 we have, renumbering the states

$$p^0(i, N|i_{-1}, N-1; \ldots; i_{-N}, 0) = p^0(i, 0|i_{-1}, -1; \ldots; i_{-N}, -N) \equiv p^0(i, 0|\mathcal{F}_N). \quad (31)$$

where $\mathcal{F}_N$ is the σ-algebra generated by $i_{-1}, \ldots i_{-N}$. Let us write

$$\pi_N = p^0(i, 0|i_{-1}, -1; \ldots; i_{-N}, -N) = p^0(i, 0|\mathcal{F}_N). \quad (32)$$

We have, because $\mathcal{F}_{N-1} \subset \mathcal{F}_N$

$$\langle \pi_N|\mathcal{F}_{N-1}\rangle = \langle p^0(i, 0|\mathcal{F}_N)|\mathcal{F}_{N-1}\rangle = p^0(i, 0|\mathcal{F}_{N-1}) = \pi_{N-1}. \quad (33)$$

So, π_N is a martingale on the σ-algebra $\mathcal{F}_N$, and by the convergence theorem, it converges almost surely to a.
limit π when $N \to \infty$.
Now if $N > n$, let us write $m = N\text{-}n > 0$. Because of the stationarity of p^0

$$p^0(i, n+m|i_{-1}, n+m-1; \ldots; i_{-m}, m) = p^0(i, 0|i_{-1}, -1; \ldots; i_{-n}, -n) = \pi_n(i). \quad (34)$$

Thus, for any fixed, positive m

$$\pi_{n+m} - \pi_n \overset{a.s.}{\to} 0. \quad (35)$$

The absolute value distance between π_{n+m} and π_n is obtained by summing $|\pi_{n+m}(i) - \pi_n(i)|$ over the M possible states i, So

$$d_{0,\ldots n+m-1}\left(q_{n+m}, q_{n+m}^{(n)}\right) = d(\pi_{m+n}, \pi_n) \overset{a.s.}{\to} 0 \text{ if } n \to \infty. \quad (36)$$

which is (29), one of our main, formal results.

2.6 *n*-times Markov approximation of the mesoscopic stationary process

Returning to inequalities (19), when the value ε is fixed for obtaining a required precision, the value $n = n(\varepsilon)$ is determined and a satisfying approximation of the exact mesoscopic process is obtained by neglecting the memory effects at time differences larger than n [18] Thus, one replaces $p(i_N, N|\, i_{N-1}, N-1; \ldots\ldots i_0, 0)$ by

$$p^{(n)}(i_N, N|i_{N-1}, N-1; \ldots\ldots i_0, 0) = p(i_N, N|i_{N-1}, N-1; \ldots\ldots i_{N-n}, N-n) \quad \text{if } N > n \quad (37)$$

With the convention

$$p^{(n)}(i_0, 0; \ldots i_N, N) = p(i_0, 0; \ldots i_N, N) \quad \text{if } N \leq n. \quad (38)$$

all the probabilities related to the approximate process $p^{(n)}$ are defined from the probabilities of p: this defines $p^{(n)}$, the approximate process of order n of p. So, $p^{(n)}$ has a finite memory of size n, whereas p has in general an infinite memory.

The process $p^{(n)}$ is a Markov process on the partial trajectories I_K consisting of groups of n successive mesocopic states

$$I_K = \left(i_{Kn}, i_{Kn+1}, \dots i_{(K+1)n-1}\right) \in \mathcal{M}^n \tag{39}$$

Its probability distributions can be written in abbreviated notations

$$P_K^{(n)}(I_0, T_0; I_1, T_1; \dots I_{K-1}, T_{K-1}) \equiv p^{(n)}(I_0, 0, 1, \dots n-1; I_1, n, n+1, \dots 2n-1; \dots; I_{K-1}, (K-1)n, \dots, Kn-1). \tag{40}$$

T_K being the group of n successive times: $T_k = \text{kn}, \text{kn}+1, \dots, (k+1)n-1$. From the approximation (18) it follows (see Appendix A) that

$$P^0\left(I_K, T_K \left| I_{K-1}, T_{K-1}; \dots I_0, T_0\right.\right) \approx P^{0(n)}\left(I_K, T_K \left| I_{K-1}, T_{K-1}\right.\right) . \tag{41}$$

where we now use the upper index^0 in P^0 and $P^{0(n)}$ to recall that, in the present section, p is the stationary distribution. Note that, because of this stationarity

$$P^{0(n)}\left(I_K, T_K \left| I_{K-1}, T_{K-1}\right.\right) = P^{0(n)}\left(I_K, T_1 \left| I_{K-1}, T_0\right.\right) = P^0\left(I_K, T_1 \left| I_{K-1}, T_0\right.\right) \equiv W(I_K | I_{K-1}). \tag{42}$$

So, the transition matrix W is well defined from the known stationary distribution p^0.

From the approximate relation (41) if follows that the exact stationnary process P^0 on the partial history I_K during the time interval T_K approximately obeys the n-times Markov Equation (see Section 2.7)

$$P^0(I_K, T_K) \approx \sum\nolimits_{I_{K-1}} W(I_K | I_{K-1})\ P^0(I_{K-1}, T_{K-1}). \tag{43}$$

while the nth approximation $P^{\,0(n)}$ satisfies (33) exactly.

2.7 Markov approximations of the nonstationary mesoscopic process

We return to the nonstationary process p generated by the deterministic microscopic process from an arbitrary initial distribution of the mesoscopic states, given by (11). As in paragraph 2.6, it is now necessary to distinguish the stationary process p° by the upper index^0.

One can write the trivial equality

$$p(i_N, N; \dots. i_{N+n-1}, N+n-1) = \sum_{i_{N-1}, \dots, i_0} p(i_{N+n-1}, N+n-1; \dots; i_N, N \,|\, i_{N-1}, N-1; \dots; i_0, 0) p(i_0, 0; \dots.; i_{N-1}, N-1). \tag{44}$$

We now use remark (14): the conditional probabilities, conditioned by the whole past up to time 0, are identical in the stationary and nonstationary situations. The stationary distributions p^0 can be approximated by its nth approximation $p^{0(n)}$ introduced in Section 2.6. Thus we can write

$$\begin{aligned} &p^0(i_{N+n-1}, N+n-1; \dots; i_N, N | i_{N-1}, n-1; \dots; i_0, 0) = \\ &p^0(i_{N+n-1}, N+n-1 | \, i_{N+n-2}, N+n-2; \dots; i_0, 0) p^0(i_{N+n-2}, N+n-2 | \, i_{N+n-3}, N+n-3; \dots; i_0, 0) \dots \\ &\dots\ p^0(i_N, N | \, i_{N-1}, N-1; \dots; i_0, 0) \approx p^{0(n)}(i_{N+n-1}, N+n-1; \dots; i_N, N | i_{N-1}, n-1; \dots; i_{N-n}, N-n). \end{aligned} \tag{45}$$

With (35), Eq. (34) yields the approximate n-times Markov Equation

$$p(i_N, N; \ldots ; i_{N+n-1}, N+n-1) \approx \sum_{i_{N-1}, \ldots, i_0} p^0(i_{N+n-1}, N+n-1; \ldots ; i_N, N$$

$$|i_{N-1}, n-1; \ldots ; i_{N-n}, N-n) p(i_{N-n}, N-n; \ldots ; i_{N-1}, N-1). \tag{46}$$

Taking $N = Kn$ for an integrer $K \geq 0$, using the condensed notations of § 2.6 and definition (42), Eq. (46) yields an approximate Master Equation for the probability $P(I_K, T)$ of the partial history I_K during the time interval T_K

$$P(I_K, T_K) \approx \sum\nolimits_{I_K} W(I_K | I_{K-1}) \; P(I_{K-1}, T_{K-1}) \tag{47}$$

which is the Eq. (43) obtained for the stationary probability $P^0(I_K, T)$. Let $P^{(n)}(I_K, T)$be the exact solution of Eq. (47) that coincides with the exact P at the n first elementary times 0, 1, ... n-1 of the system history: $P^{(n)}(I_0, T) = P(I_0, T)$. Then, $P^{(n)}(I_K, T)$ defines the nth approximation of $P(I_K, T)$: in principle, it can be computed from Eq. (47) since the probability transitions W are known by (41).

The stationary probabilities approximation $P^{0(n)}$ deduced from p^0 provide the stationary solution of (47)

$$\begin{aligned} P^{0(n)}\left(I_K, T_K\right) &= p^0\left(i_{Kn}, Kn; \ldots ; .i_{K(n+1)-1}, K(n+1)-1\right) \\ &= p^0\left(i_{Kn}, 0; \ldots ; i_{K(n+1)-1}, n-1\right). \end{aligned} \tag{48}$$

So, when $K \to \infty$,

$$P^{(n)}\left(I_K, T_K\right) \to P^{0(n)}\left(I_K, T_K\right). \tag{49}$$

and consequently, for any integer $k \in [0, n\text{-}1]$, the nth approximation of the mesoscopic distribution p satisfies

$$p^{(n)}(i, Kn+k) \to \mu(i,k) = \mu(i) \;\text{ if } K \to \infty. \tag{50}$$

for any initial mesocopic distribution, which is the basic assumption of statistical thermodynamics. Supplementary assumptions allow one to conclude that, in realistic situations, the mesoscopic distribution p itself satisfies this property (see Appendix B).

2.8 Time averages and simple Markov approximation

Up to now, we took as time unit some time step τ which gives the time scale of microscopic phenomena. By considering some finite partition (i) of the phase space X and replacing the microscopic states $x \in X$ by the mesocopic states $i \in (i^k)$, we have performed a space coarse graining, as necessary for taking practical observations into account. For the same purpose, one should also introduce [18] a space coarse graining, since the time scale $\theta = n\,\tau$ of current observations is much larger than τ: $n >> 1$.

All mesoscopic functions remaining practically constant on the time scale θ, their averages can be computed from the time averages $\overline{P}_K$ of the probabilities p_k over θ

$$\overline{p}_K = \frac{1}{n} \sum\nolimits_{k \in T_K} p_k. \tag{51}$$

where K is an integer ≥ 1 and T_K is the time interval (τ being the time unit) T_K = $(K\text{-}1)n$, $K + 1$, ... Kn-1.

Suppose (a) that the mesoscopic probabilities p are slowly variating functions of the mesoscopic states, (*i.e.* for any positive α, $|p(i) - p(j)| < \alpha$ if the distance between the mesostates i and j is small enough, with an appropriate metric in the spase of mesostates), and (b) that discontinuous trajectories have low probabilities and can be neglected. Of course, these assumptions are not verified for some important, well known processes such as Brownian processes, but they seem to be reasonable for modeling physical processes where the inertial effects are strong enough. Then, a simple approximation is to consider that

$$\begin{aligned} &p^0\left(i_{Kn-1}, Kn-1; \ldots; i_{(K-1)n}, (K-1)n \,|i_{(K-1)n-1}, (K-1)n-1; \ldots; i_{(K-2)n}, (K-2)n\right) \\ &\approx p^0\left(i_{\overline{K}}, Kn-1; \ldots; i_{\overline{K}}, (K-1)n \,|\, i_{\overline{K}-1}, (K-1)n-1; \ldots; i_{\overline{K}-1}, (K-2)n\right) \\ &\equiv \overline{W}\left(i_{\overline{K}} | i_{\overline{K}-1}\right). \end{aligned} \tag{52}$$

where

$$\overline{K} = \frac{1}{n} \sum\nolimits_{k \in T_K} k = \frac{1}{n} \sum\nolimits_{k=(K-1)n}^{Kn-1}. \tag{53}$$

Consider the time-averaged probability

$$\overline{P}(i_{\overline{K}}, K) \equiv \frac{1}{n} \sum\nolimits_{k \in T_K} p(i_k, k) \approx \frac{1}{n} \sum\nolimits_{k \in T_K} p\left(i_{\overline{K}}, k\right). \tag{54}$$

Using the Markov Eq. (47) and the complementary approximations (42), we obtain the new Master Equation

$$\overline{P}(i, K) \approx \sum\nolimits_j \overline{W}(i\,|j)\; \overline{P}(j, K-1). \tag{55}$$

This equation is much simpler than Eq. (47), since it applies in the space $\mathcal{M}$ of the M mesostates (i), whereas (47) is valid in the space $\mathcal{M}^n$ of n successive mesostates. However, Eq. (45) relies on several approximations that are difficult to control. In spite of these difficulties, which can only be precisely discussed for specific examples, Master Equations like (55), resulting from deterministic microscopic systems by coarse-graining both their states and time, are a practical way to study their evolution of a mesoscopic scale, used in innumerable works.

3. Discussion of the Markov representation derived from Hamiltonian dynamics, and estimation of the uniformization time

The previous results show that the coarse grained mesoscopic dynamics can eventually be represented by a Master Equation, because the memory of this dynamics is gradually lost over time. However, they do not provide the time scale of this fading. In order to estimate its order of magnitude simply, we make an intuitive remark: the conditional probability to jump from some mesostate i to another one can be evaluated without knowing the past history of the system if one knows the initial microscopic distribution over i. The only unbiaised initial distribution is the uniform one. Thus, one can consider that the system has a memory limited to one time step if uniformity is approximately realized in each mesoscopic cell: this is the basis of the elementary Markov models of mesoscopic evolution. Let T be the average time needed to reach uniformity in a mesoscopic scale, starting from strong inhomogeneity. In a first approximation it is reasonable to use this uniformization time T to characterize the time scale over which a Markov evolution can describe the system.

3.1 Uniformization time in a mesoscopic cell: An elementary estimation for Hamiltonian systems

Using oversimplified, but reasonable arguments [17], we now coarsely estimate the uniformization time T in a mesoscopic cell. As an example, we consider n identical particles initially located in this cell, among N identical particles in an isolated vessel. The complete system obeys Hamilton mechanics.

Assume that the particles constitute a gas under normal conditions, with density $\rho \approx 3.\ 10^{25}$ molecules.m^{-3}. A mesocopic state can be reasonably represented by a cube of size $l \approx 10^{-6}$ m (as an order of magnitude), which contains $n \approx 3.\ 10^{7}$ molecules. We now divide the mesoscopic cell into m "microscopic" cells whose size λ is comparable to the size of a molecule: each of these microscopic cells, however, should contain a sufficient number particles for allowing them to interact from time to time. We can take $\lambda \approx 10^{-8}$ m, so each microscopic cell approximately contains 30 molecules, and there are $m \approx 10^{6}$ microscopic cells in a mesocopic cell. The particles have an average absolute value $v \approx 500$ m.s^{-1} in typical conditions. They can jump between the various microcells of the same mesocopic cell. They can also jump out of their initial mesoscopic cell, but they are replaced by molecules proceeding from other cells, and we assume that these contrary effects coarsely compensate themselves, except in the first stage of the evolution if the initial mesocopic distribution is strongly inhomogeneous.

Because all particles are identical, an almost microscopic configuration of a mesoscopic cell can be defined by specifying the number of particles in each of its microscopic cells. Focusing on a given mesoscopic cell, we compute the number of its possible configurations, and we estimate the average time θ necessary for the system to visit all these configurations. Note that the uniformization time T is obviously much larger than θ: $T >> \theta$. So, θ is a lower bound of T.

The number of ways of partitioning the n identical particles into the m microscopic cells is

$$C = \frac{(m+n-1)!}{n!(m-1)!} \approx \exp\left[(m+n)\,\varphi(x)\right] \text{ with } \varphi(x) \\ = -x\ \ln x - (1-x)\ \ln(1-x) \text{ and } x = n/(m+n). \tag{56}$$

The system jumps from one of these configurations to another one each time one the present particles jumps to another microscopic cell. The order of magnitude of the time needed for a particle to cross a micro-cell is λ/v, and the time between two configurations changes is $\tau \approx (1/n)\ \lambda/v$.. In order that all configurations are visited during time θ we should have at least $\theta \approx C\,\tau$ (in fact, θ should be much larger than $C\tau$ because of the multiple visits during θ). So we conclude from (46) and relevant approximations that a lower bound of Θ satisfies

$$n\frac{\varphi(x)}{x} \approx \ln\frac{v\Theta}{\lambda} \quad \text{with } x = n/(m+n) \approx 1. \tag{57}$$

With the previous numerical values

$$\theta \approx \frac{\lambda}{v}\left(\frac{n}{m}\right)^{m} \approx 2.10^{-11}.(30)^{10^6}\ \text{s}. \tag{58}$$

which is far larger than the age of universe (now estimated to be about 14×10^9 years, or 4.4×10^{17} s)!

Although these calculations are very rudimentary, it is clear that, in the framework of purely Hamiltonian systems, the microscopic distribution within a mesocopic cell remains far from uniformity during any realistic time if it is initially fairly inhomogeneous.

More generally, it is clear that the uniformization time T should be of the order of Poincaré time [32–36] in a mesoscopic cell, which is known to be extraordinarily long [9, 37].

3.2 An elementary, empirical approach of mesoscopic systems

The practical relevance of Markov processes to model a large class of physical systems is supported by a vast literature. We have seen that the progressive erasure of its memory over time allows one to justify the use of a Markov process to represents the evolution of the coarse-grained system. However, such representation can also stem from random disturbances due to the measurements or other sources of stochasticity: then, one has to renounce to a purely deterministic microscopic dynamics, as formerly proposed by many authors, even without adopting the formalism of Quantum Mechanics. It is interesting to compare the time scales of the relaxation to equilibrium in both approaches with an elementary example.

3.2.1 Uniformization induced by randomization

Suppose now that the measure process does not induce any significant change in the average molecules energy - so, their average velocity remains unchanged – but that it causes a random reorientation of their velocity. A rudimentary, one dimensional model of such a randomization could be to assume that each time a molecule is about to pass to a neighboring cell, it will go indifferently to one of the neighboring microscopic cell. In a one dimensional version of the model, a molecule perform a random walk on the $\eta = l/\lambda = 10^2$ points representing the microscopic cells contained in the mesoscopic cell, and we adopt periodic conditions at the boundaries of the mesocopic cell. The $\eta \times \eta$ transition matrix of the process is a circulant matrix which, in its simplest version, has transition probabilities ½ to jump from any state to one of its neighbors, and it is known that its eigenvalues λ_k are $\lambda_k = \cos(2\pi k/\eta)$, $k = 0, 1, \ldots [\eta/2]$. The number of jumps necessary for relaxing to the uniform, asymptotic distribution is of the order

$$1/(-\ln \lambda_1) \propto 2(2\pi/\eta)^{-2} \approx 500 \text{ s}.$$

which correspond to a relaxation time of $500.\ \lambda/v \approx 10^{-8}$ s, which is very short for current measurements, but comparable with (or even larger than) the time scale of fast modern experiments. Considering a 3-dim model would not change this time scale significantly. It is conceivable that he molecules are not necessarily reoriented each time they leave a microscopic cell. Even if the proportion of reoriented molecules is as low as 10^{-6}, the relaxation time is of order 10^{-2} s, which is insignificant in many simple measures. In this case the Markov representation can be justified.

3.2.2 Semi-classical Hamiltonian systems

In analogy with the previous randomized system, we can introduce a new source of stochasticity in the coarse-grained deterministc systems considered in Sections 2 and 3. This could be done by assuming that a particle cannot be described by a point, but by a probability density centered on the point that would represent it classically: such a description borrows one, but not all, of the axioms of wave mechanics, and it can be qualified as a "semi-quantical" description. A similar assumption can be introduced without referring to quantum mechanics, by noticing that a particle cannot be localized in a given mesoscopic cell with complete certainty, because of its finite size: if it is mainly attributed to a given cell, there exists a small probability that it also belongs to a neighboring cell. Even without formalizing these possibilities, one can presume that such random effects shorten drastically the memory of the mesoscopic process, and make it short with respect to ordinary measure times: then the Markov approximation described in Section 2 can correctly represent the evolution of the observed coarse-grained process.

4. Conclusion

We have studied the mesoscopic, stochastic process derived from a deterministic dynamics applied to the cells determined by measure inaccuracies. The stationary process, which arises when the microscopic initial state is distributed according to a time invariant measure, was studied by Kolmogorov and further authors: we extended their methods and some of their results, and considered the nonstationary process which stems from a noninvariant initial measure. We have shown that, according to Jaynes' principle, the "exact" mesocopic process can be approximately replaced by the Markov process which, at any time n, reproduces the one-time probability of each mesostate and the transition probabilities from it. This Markov process maximizes the trajectory entropy up to time n, as well as the entropy at time n, conditioned by prior events. The Jaynes' principle, however, does not control the accuracy of this estimate: this was our next concern.

So, a sequence of successive approximations has been defined for the stationary mesoscopic process, based on one of our main results: the probability of any mesostate state conditioned by all past events, can be approximated by its probability conditioned by the n last past events only, the integer n being determined by the maximum distance allowed between these probabilities, as small as it may be. This property entails that the nonstationary mesoscopic process can be approximated by a n-times Markov process or even, after a time coarse-graining, by an ordinary one-time Markov process. These approximations require certain conditions which should be fulfilled by "normal" physical systems, with possible exceptions for slowly relaxing systems. If they are satisfied, the existence of a thermodynamic equilibrium is derived for a coarse-grained system obeying a measure-preserving deterministic dynamics, in particular an Hamiltonian dynamics, without introducing *ad-hoc* external noises. However, very rough estimations of the relaxation time show that for reasonable values of the parameters this time is extraordinarily long and completely unrealistic.

We conclude that, although the basic hypotheses of thermodynamics can be justified from a Hamiltonian or deterministic microscopic dynamics applied to the mesoscopic cells, the observed time scales of the relaxation to equilibrium cannot be explained without going beyond pure Hamilton mechanics, by introducing additional random effects, in particular due to the intrinsic imprecision of the particles localization.

Appendix A: Approximating the n-times conditional probability

With the notations of Section 4.2, we consider approximation (55), which is the basis of the n-times Markov approximation both in the stationary and nonstationary situations. Repeating approximation (51) we can write

$$p(i_{N+n-1}, N+n-1; \dots; i_N, N \,|\, i_{N-1}, N-1; \dots; i_{N-n}, N-n; \dots; i_0, 0) =$$
$$= p(i_{N+n-1}, N+n-1 | i_{N+n-2}, N+n-2; \dots; .i_0, 0)\dots\dots p(i_N, N | i_{N-1}, N-1; \dots; i_0, 0)$$
$$\approx p^{(n)}(i_{N+n-1}, N+n-1 | i_{N+n-2}; \dots i_{N-1}, N-1)\dots p^{(n)}(i_N, N | i_{N-1}, N-1; \dots; i_{N-n}, N-n). \tag{59}$$

The last line of (49) is $p^{(n)}(i_{N+n-1}, N+n-1; \dots; i_N, N \,|\, i_{N-1}, N-1; \dots; i_0, 0)$. We write

$$p(i_{N+n-1}, N+n-1; \dots; i_N, N \,|\, i_{N-1}, N-1; \dots; i_{N-n}, N-n; \dots; i_0, 0) \equiv$$
$$p^{(n)}\left(i_{N+n-1}, N+n-1; \dots; i_N, N \,|\, i_{N-1}, N-1; \dots; i_{N-n}, N-n; \dots; i_0, 0\right) Q_N^{(n)}. \tag{60}$$

and for $k > n$ we define l_k, using the abbreviations (32)

$$l_k(i_0, \dots i_{k-1}) \equiv \ln \frac{\Pi_k(i_k)}{\Pi_k^{(n)}(i_k)}. \tag{61}$$

We have by (24)

$$s_n(p) - s_k(p) = \sum\nolimits_{i_0, \dots i_k} p_N(i_0, 0; \dots; i_k, k) \ln \frac{\Pi_k(i_k)}{\Pi_k^{(n)}(i_k)} = \langle l_k(i_0, \dots i_{k-1}) \rangle \equiv \sigma_k(n). \tag{62}$$

(Note that σ_k is positive, although this not necessarily true for l_k). By (24) for any positive ε

$$2\left\langle d^2\left(p_k, p_k^{(n)}\right)\right\rangle \leq \langle \sigma_k(i_0, \dots i_{k-1}) \rangle < \varepsilon \text{ if } n \text{ is large enough.} \tag{63}$$

Averaging the logarithm of Eq. (60) we have

$$\left\langle L_N^{(n)} \right\rangle \equiv \left\langle \ln Q_N^{(n)} \right\rangle = \sum\nolimits_{k=0}^{n-1} [s_n(p) - s_{N+k}(p)] \propto n\, [s_n(p) - s_\infty(p)] = n\, \delta s(n). \tag{64}$$

$\delta s(n) \equiv s_n(p) - s_\infty(p)$ can be interpreted as an entropy fluctuation with respect to its equilibrium thermodynamic value. If such a fluctuation relaxes exponentially to 0 with time, as usual, the last term of (54) tends to 0 when $n \to \infty$. Then, the n-times Markov approximations 4.2 and 5.1 are justified. Although exponential relaxation can be considered as a characteristic of "normal" physical systems, slower relaxations can occur: in this case the Markov approximation may be invalid.

Appendix B: Tendency to the stationary mesoscopic distribution

This tendency can be reasonably expected from the approximation of the exact mesocopic process by Markov processes, but it can only be affirmed by adding additional assumptions to the basic assumptions. We first prove a simple, useful lemma.

B.1. Lemma. Consider a d-dim sequence $u_{n,\mathrm{k}}$ with 2 positive, integer indices n, k, satisfying the following properties:

i. it is absolutely bounded: there is a positive real number M such that $|u_{n,k}| < M$ for all integers n, k.

ii. for all n, k, there are positive numbers ε_n (independent of n) and ν (independent of n and k) such that

$$|u_{n,k}| < \varepsilon_n + \nu\, |u_{n,k-1}| \text{ and } \varepsilon_n \to 0 \text{ if } n \to \infty. \tag{65}$$

Then $u_{n,k} \to 0$ if $n \to \infty$ and $k \to \infty$.
In fact, for any positive ε, there is an integer n_0 such that $\varepsilon_n < \varepsilon$ if $n > n_0$, and

$$|u_{n,k}| < \varepsilon\left(1 + \nu + \ldots + \nu^{k-1}\right) + \nu^k u_{n,0} < \frac{\varepsilon}{1-\nu} + \nu^k\, |u_{n,0}| \text{ if } n > n_0. \tag{66}$$

So, $|u_{n,k}|$ can be made as small as desired by chosing n and k large enough.

B.2. For given integers n and K larger than 1, and states $i_k \in$ M, k = 0, 1, ... $(K + 1)n$-1, we will write.

$p(i_0, 0; i_1, 1; ..; i_{(K+1)n-1}, (K+1)n - 1) \equiv p(0, 1, \ldots, (K+1)n - 1)$ for the sake of simplicity. In these abbreviated notations, we have

$$p(Kn, Kn+1; .., (K+1)n - 1) = \sum_{i_0, i_1, \ldots\, i_{Kn-1}} p((K+1)n, -1 \ldots Kn | Kn - 1, \ldots 0)\; p(0;\ 1; .., Kn - 1). \tag{67}$$

We know that

$$p((K+1)n, -1 \ldots Kn | Kn - 1, \ldots 0) = p^0((K+1)n, -1 \ldots Kn | Kn - 1, \ldots 0)\,. \tag{68}$$

p^0 being the stationary probabilities, and that, for large n

$$p^0((K+1)n, -1 \ldots Kn | Kn - 1, \ldots 0) \approx p^0((K+1)n, -1 \ldots Kn | Kn - 1, \ldots K(n-1)). \tag{69}$$

More precisely, in the conditions discussed previously, for any given positive ε, there is an integer $n(\varepsilon)$ such that

$$|p^0((K+1)n, -1 \ldots Kn | Kn - 1, \ldots 0) - p^0((K+1)n, -1 \ldots Kn | Kn - 1, \ldots K(n-1))| < \varepsilon \quad \text{if } n \geq n(\varepsilon) \tag{70}$$

So, Eq. (67) becomes

$$\begin{aligned} p(Kn, Kn+1; .., (K+1)n - 1) = {} & \\ \sum_{i_0, \ldots\, i_{Kn-1}} & [p((K+1)n, -1 \ldots Kn | Kn - 1, \ldots 0) - p^0((K+1)n, -1 \ldots Kn | Kn - 1, \ldots 0)]\, p(0;\ 1; .., Kn - 1) \\ + \sum_{i_0, \ldots\, i_{Kn-1}} & p^0((K+1)n, -1 \ldots Kn | Kn - 1, \ldots (K-1)n)\, p(K-1)n; .., Kn - 1) \\ \equiv a_{n,K} + {} & \sum_{i_0, \ldots\, i_{Kn-1}} p^0((K+1)n, -1 \ldots Kn | Kn - 1, \ldots (K-1)n)\, p(K-1)n; .., Kn - 1). \end{aligned} \tag{71}$$

where the 1st term of the last line satisfies $|a\ (n, K)| < \varepsilon$ if $n \geq n(e)$.

The second term in the last line of (71) is, in other notations, the righth hand side term of the approximate Master Eq. (47) of Section 2.8

$$P(I_K, T_K) \approx \sum\nolimits_{I_K} W(I_K \, |I_{K-1}) \; P(I_{K-1}, T_{K-1}). \tag{72}$$

This approximate Master Equation can now be written more precisely, in the notations of 2.8

$$P(I_K, T_K) = \; A_{n,K}(I_K, T_k) + \; \sum\nolimits_{I_K} W\left(I_K, T_K \, \middle|I_{K-1}, T_{K-1}\right) \; P(I_{K-1}, T_{K-1}) \; . \tag{73}$$

where $A_{n,K}(I_K, T_k)$ is just the $a_{n,K}$ term of (71) expressed in the notations of 2.8 where T_K is the group of n successive times: $kn, kn+1, .., (k+1)n-1$, and $I_K = \left(i_{Kn}, i_{Kn+1}, \ldots i_{(K+1)n-1}\right) \; \in \; \mathcal{M}^n$ describes the corresponding partial history of the mesoscopic system.

On the other hand, we know that the stationary distribution P^0 satisfies the Master Eq. (72) exactly. So, writing

$$U_{n,K}(I_K, T_K) \; \equiv \; P(I_K, T_K) - P^0(I_K, T_K). \tag{74}$$

we have

$$U_{n,K}(I_K, T_K) = \; A_{n,K}(I_K, T_k) + \; \sum\nolimits_{I_K} W(I_K \, |I_{K-1}) \; U_{n,K-1}(I_{K-1}, T_{K-1}). \tag{75}$$

Note that W depends of n, but is independent of K.

From (72)–(75) it results that $U_{n,K}(I_K, T_K)$ tend to 0 when n and K tend to infinite if, furthermore, the following.

condition (c) holds. In fact, the (M^n- dim) vector $U_{n,K}$ is orthogonal to the left-eigenstate with eigenvalue 1.

of matrix W. All the eigenvalues of the projection of W in the corresponding subspace have an absolute value smaller than 1. Thus the lemma 1 applies if condition (c) is satisfied:

(c) When n increases, the absolute values of the nonstationary eigenvalues of W have an upper bound <1.

This property is likely to hold if the actual stationary mesoscopic process is not too different from an exact Markov process. So, it is reasonable to conjecture that property (c) holds for typical actual systems.

Author details

Michel Moreau* and Bernard Gaveau
Sorbonne Université, Paris, France

*Address all correspondence to: mlmoreau@free.fr

References

[1] Boltzmann L. Vorlesungen über Gastheorie. In: *part*. Vol. *2*. Leipzig: J. A. Barth; 1898

[2] L. Boltzmann. Reply to Zermelo's Remarks on the theory of heat.In: History ofmodern physical sciences: The kinetic theory of gases, ed. S. Brush, ImperialCollege Press, **57**, 567 (1896).

[3] Ehrenfest P, Ehrenfest T. *The conceptual foundations of the statisticalapproach in Mechanics*. New York: Dover; 1990

[4] Uhlenbeck GE. *Anoutline of Statistical Mechanics, in Fundamental problemsin Statistical Mechanics*. II. ed. Cohen, North Holland, Amsterdam: E. G. D; 1968

[5] Landau LD, Lifshitz EM. *Statistical Physics*. 3rd ed. Oxford: Pergamon Press; 1980

[6] Landau LD, Pitaevskii LP. *Physical Kinetics*. Oxford: Pergamon Press; 1981

[7] H.B. Callen, *Thermodynamics and an Introduction to Thermostatistics* (John Wiley and Sons, New York, (1985).

[8] Reif F. *Fundamentals of Statistical and Thermal Physics*. NY: Mc Graw Hill; 1965

[9] J.P. Eckmann and D. Ruelle, *Ergodic theory of chaos and strange attractors*, Rev Mod Phys**57**, 617 (1985).

[10] J.R. Dorfman, A, Introduction to Chaos I, Nonequilibrium Statistical Mechanics (Cambridge, New York, (1999).

[11] Gallavotti G. Statistical Mechanics: a Short Treatise. Berlin: Springer; 1999

[12] Gallavotti G. Journal of Statistical Physics. 1995;**78**:1571

[13] Evans DJ, Cohen EGD, Morries GP. Phs. Rev. Lett. 1993;**71**:15

[14] Gallavotti G, Cohen EGD. Phys. Rev. Lett. 1995;**74**:2694

[15] Yvon J. *La Théorie Statistique des Liquides*. Paris: Herman; 1935

[16] Forster D. *Hydrodynamic Fluctuations*. Benjamen, Reading, Masschusetts: *Broken Symmetry and Correlation Function*s; 1975

[17] B. Gaveau and L.S. Schulman, EPJST **224**, 8, p 891 (2015).

[18] Gaveau B, Moreau M. Chaos. 2020; **30**:083104

[19] Arnold VI, Avez A. *Ergodic probems of Classical Mechanics*. Mathematical Physics Monographs: Benjamin; 1968

[20] Doob J. *Stochastic Processes*. N.Y: Wiley; 1953

[21] Levy P. *Théorie de l'addition des variables aléatoires*. Paris: Gauthier-Villars; 1937

[22] Feller W. *An Introduction to Probability Theory and Its Applications*, vol II. N.Y: Wiley; 1971

[23] McKean HP. *Stochastic Integrals*. London: Academic Press, NY; 1968

[24] Gaveau B, Gaveau MA. *Diffusion of a particle in a very rarefied gas*, Symposium on rarefied gas dynamics, Muntz, Weaver, Campell eds. Progress in Astronomics and Aeronautics. 1988; **118**(61)

[25] Jaynes ET. Phys. Rev. II **108**, 171 (1957). Phys. Rev. II. 1957;**106**:620

[26] Khinchin AI. *Mathemtical Foundations of Statistical Mechanics*. N.Y: Dover; 1949

[27] C. E. Shannon, *A mathematical theory of communication*, Bell System

Technical Journal, vol. 27, p. 449–423 and 623–656 (1948).

[28] Brillouin L. *Science an Information Theory*. NY: Academic Press; 1956

[29] Khinchin AI. *Mathematical Foundations of Information Theory*. N.Y: Dover; 1957

[30] MacMillan R. *The basic theorems of information theory*. Ann Math Statistics. 1953;**24**:193

[31] T. Cover. and J. Thomas,*Elements of Information Theory*. Wiley: N.Y; 1991

[32] H. Poincaré. Acta Math. **13**, 1 (10890).

[33] Kac M. Bull. Amer.Math. Soc. 1947; **53**:1002

[34] Wolfowitz J. Bull. Amer. Math. Soc. 1949;**55**:394

[35] Blum JR, Rosenblatt JI. J. Math. Sci. (Delhi). 1967;**2**:1

[36] Schulman LS. Phys. Rev. A. *1978;**18*** (5):*2379*

[37] Gallavotti G, Cohen EGD. J. Stat. Phys. 1995;**80**:931

Chapter 6

Qualitative Analysis for Controllable Dynamical Systems: Stability with Control Lyapunov Functions

Adela Ionescu

Abstract

The present chapter focuses on some recent work on the qualitatively analysis of dynamical systems, namely stability, a powerful tool with multiple connected appliances. Among them, feedback is a powerful idea which is used extensively in natural and technological systems. In engineering, feedback has been rediscovered and patented many times in many different contexts. Stabilizing a dynamical system could be often easier if we approach *controllable systems*. When the dynamical system is in a controllable form, we can place bounds on its behavior by analyzing the improvement of the linear and nonlinear operators that describe the system. In this chapter it is analyzed how a control in a simple form, could influence the possibility to construct the so-called *Control Lyapunov Function (CLF)* in order to stabilize the dynamical system in study. The main idea is to test multiple cases, in order to get a rich information panel and to make easier the problem of finding a CLF, which is generally a difficult task. As applications, models from excitable media are chosen.

Keywords: dynamical systems in control, Lyapunov stabilities, stabilization by feedback, Lyapunov and storage function, feedback control, computational methods, algebraic methods

1. Introduction: general outlines in feedback and stability for dynamical systems

The term *feedback* is used to refer to a situation in which two (or more) dynamical systems are connected together such that each system influences the other and their dynamics are thus strongly coupled. Feedback is a powerful idea whose principle is based on corrections on the difference between desired and current performance. It was applied, rediscovered and patented in different contexts in engineering. The feedback methods had important improvements in time and, due to its remarkable properties, these improvements had significant importance in all applied sciences models.

A basic feature of feedback is that it changes the dynamics of a system. By modifying the behavior in the sense needed by the application, we can stabilize the model which is initially unstable, or we can obtain responsive systems from sluggish ones.

A survey on control process strategies and applications shows that: (1) a variety of nonlinear controller design techniques are based on input–output linearization; (2) few experimental studies of these techniques have been presented; and (3) many important problems remain unsolved [1].

1.1 Feedback model outlines

There are several types of finite-dimensional, nonlinear process models. The *continuous-time, state-space* model has the form:

$$\dot{x} = f(x) + g(x)u \qquad (1)$$
$$y = h(x).$$

x is the vector of state variables, $x \in R^n$, **u** is the vector of input variables, **y** the vector of controlled output variables, $u, y \in R^m$; **f** and **h** are vectors of nonlinear functions, $f \in R^n, h \in R^m$; finally **g** is a matrix of nonlinear functions.

The single-input, single-output (SISO) case where $m = 1$ is generally easiest and good to facilitate understanding the basic concepts. Consider the Jacobian linearization of the nonlinear model

$$\dot{x} = \left[\frac{\partial f(x_0}{\partial x} + \frac{\partial g(x_0}{\partial x} u_0\right](x - x_0) + g(x_0)(u - u_0) \qquad (2)$$
$$y - y_0 = \frac{\partial h(x_0)}{\partial x}(x - x_0).$$

Using derivation variables, the Jacobian model can be written as a linear state-space system

$$\dot{x} = Ax + Bu \qquad (3)$$
$$y = Cx$$

with obvious definitions for the matrices A, B and C. It is important to note that the Jacobian model is an exact representation of the nonlinear model only at the point (x_0, y_0). As result, a control strategy based on a linearized model may involve unsatisfactory performance and robustness at other operating points.

Roughly speaking, *feedback linearization* is a collection of ways for transforming the original system models into equivalent models of a simpler form. The central idea of feedback linearization is to algebraically transform nonlinear systems dynamics into (fully or partly) linear ones, in order to enable to apply linear control techniques. This approach is essential different from the classic Jacobian linearization, because feedback linearization is realized by an exact state transformation and a feedback law, rather than by linear approximations of the dynamics of the model. More important is the *local* feedback linearization, as it allows avoiding complications associated with the global problem.

After feedback linearization, the input–output model is linear:

$$\dot{\xi} = A\xi + Bv \qquad (4)$$
$$w = C\xi.$$

Here ξ is the vector of transformed variables, $\xi \epsilon R^r$; v is the vector of input variables; w is the vector of output variables, $w, v \epsilon R^m$, and A, B and C are matrices with simple canonical structure. The integer r is the fundamental characteristic of a nonlinear system, named the *relative degree*; if $r < n$, we have to complete the coordinate transformation by adding additional n-r state variables [1].

For a dynamical system, the *controllability* problem is to check the existence of a forcing term or control function u(t) such that the corresponding solution of the system will pass through a desired point, $x(t_*) = x_*$. The initial form of the controlled system motivates the form of the control. Thus, a controlled system will have a complex dynamic and therefore, analyzing its stability implies analyzing the nonlinear operators that describe the system's components. In applied sciences and engineering models, the control is aimed to compare the system against the desired behavior and compute corrective actions based on a model of the system's response to external inputs. Therefore, the modern control techniques include the use of algorithms [1].

1.2 Stability outlines

Let us consider the solution of a differential equation representing a physical phenomenon or the evolution of some system. There always is some uncertainty concerning the initial conditions, because, when one attempts to repeat a given experiment, the reproduction of the initial conditions is never entirely identical. It is thus fundamental to be able to recognize the circumstances under which small variations in the initial conditions will only introduce small variations in what follows of the phenomenon.

It is known that stability is a property of the solutions of differential equations in R^n of the form $\dot{\boldsymbol{x}} = \boldsymbol{f}(t, \boldsymbol{x})$ by which, given a "reference" solution $x^*\left(t, t_0^*, x_0^*\right)$, any other solution $x(t, t_0, x_0)$ starting close to $x^*\left(t, t_0^*, x_0^*\right)$ remains close to $x^*\left(t, t_0^*, x_0^*\right)$ for long times. Thus, generally speaking, we can state the question of stability as: "small variations in the initial conditions will imply small variations in what follows for the phenomenon".

The stability concept has the beginnings back in the past, in the analysis of the planets motion. Then Lagrange, Dirichlet had refined the definition, including the boundedness of trajectories. But Lyapunov's work was a corner stone in this area, by analyzing the stability concept, with the help of a positive non-decreasing function which is decreasing along the system trajectories. The Lyapunov functions are a mainstay in the control theory and in the applied sciences modeling.

Within the mathematical context they are implied, the Lyapunov functions provide sufficient conditions for the stability of equilibrium and for analyzing its basin of attraction or more general invariant sets. They characterize the long-time behavior of the solutions depending on their initial solutions. Therefore it appears the natural question how to compute a Lyapunov function for a particular system? Although the existence of Lyapunov functions has been studied in few theorems, it is not provided yet a general method to compute them. The converse theorems which appeared around 1950 were a great help in the issue. A converse theorem generally establishes that, if a system has a certain kind of stability, then there exists a Lyapunov function for the system that characterizes that kind of stability. Still, the converse theorems are not very constructive in practice, since they use the solution trajectory of the system to construct the Lyapunov function and the solution trajectories are usually not known. Krasovski himself noted that [2]:

"One could hope that a method for proving the existence of a Lyapunov function might carry with it a constructive method for obtaining this function. This hope has not been realized".

Numerous computational construction methods have been developed in mathematical community, based on different methods such as linear matrix inequalities, linear programming, series expansion, algebraic methods, theoretic methods and many others.

Very important to notice is the Lyapunov theorem, which enable establishing stability or asymptotic stability of equilibrium points without explicitly computing trajectories [2].

The Lyapunov theorem is of fundamental importance in system theory. It asserts the possibility of establishing stability or asymptotic stability of equilibrium points without explicitly computing trajectories [2].

Theorem 1 (Lyapunov). Let $x_e = 0$ be an equilibrium point for the system (1). Let $V : R^n \to R$ be a positive definite continuously differentiable function.

1. If $\dot{V} : R^n \to R$ is negative semi-definite, then x_e is stable;

2. If $\dot{V}$ is negative definite, then x_e is asymptotically stable.

The theorem assesses the existence of a Lyapunov function but does not provide a method to compute one. In the case of linear systems, this issue arises naturally, but in general computing a Lyapunov function is an open problem giving rise to different ways to construct it.

We recall in what follows the two basic Lyapunov criteria.

a. *The first Lyapunov criterion* is based on the eigenvalues analysis.

Let us consider the following continuous-time nonlinear system:

$$\dot{\boldsymbol{x}} = \boldsymbol{f}(\boldsymbol{x}(t), u(t)). \tag{5}$$

In the vicinity of the equilibrium point (x_0, u_0), let us consider the corresponding linearized system:

$$\dot{\tilde{x}}(t) = A\tilde{x}(t) + B\tilde{u}(t). \tag{6}$$

This criterion has three distinct cases for the eigenvalues λ_i of the matrix A [3]:

i. If $Re\ \lambda_i < 0$ for all i, then (x_0, u_0) is asymptotically stable;

ii. If there exits at least one i such as $Re\ \lambda_i > 0$ then (x_0, u_0) is unstable;

iii. If there exits at least one i such as $Re\ \lambda_i = 0$ and for all other λ_j, $j \neq i$, $Re\ \lambda_j < 0$, then we cannot conclude anything about the stability of (x_0, u_0). In this case we say that the criterion is not effective

b. *The second Lyapunov criterion*

Theorem 2. Consider the dynamical system in R^n:

$$\dot{\boldsymbol{x}}(t) = \boldsymbol{f}(\boldsymbol{x}(t)) \tag{7}$$

and let $x = 0$ be its unique equilibrium point. If there exists a continuously differentiable function $V : R^n \to R$ such that:

$$V(0) = 0; \tag{8}$$

$$V(\boldsymbol{x}) > 0, \forall x \neq 0; \tag{9}$$

$$\|\boldsymbol{x}\| \to \infty \Rightarrow V(\boldsymbol{x}) \to \infty; \tag{10}$$

$$\dot{V}(\boldsymbol{x}) < 0 \forall x \neq 0. \tag{11}$$

Then $x = 0$ is global asymptotically stable.

The condition (9) refers to the *monotonicity* of the Lyapunov function. We say that V is decreasing along trajectories, using the *orbital derivative* given by:

$$\dot{V}(\mathbf{x}) = \left\langle \frac{\partial V(\mathbf{x})}{\partial x}, f(\mathbf{x}) \right\rangle \tag{12}$$

where $\langle , \rangle$ is the inner product in R^n and $\frac{\partial V}{\partial x}$ is the gradient of V. Also, the condition (10) refers to the requirement for V to be *radially unbounded*.

We could ask if Lyapunov functions always exist, and if so, how could we find such a function? For the first part of the question the answer is generally positive but, finding a Lyapunov function is not immediate, since the converse theorems assume the knowledge of the solutions of the system (7) [2, 3]. Therefore refining the definition of Lyapunov function and establishing a more specific context was very necessary.

An important aim in qualitative analysis of the stability is to search if the solutions remain close to the equilibrium and moreover, if they converge towards it. Therefore the search for the Lyapunov function must be more accurate. The *strict Lyapunov functions* can achieve this goal. Designed as generalization of the energy in a physical dissipative system, they preserve the property of decreasing energy along trajectories and thus, the solutions of the system converge to a (local) minimum of energy.

Definition 1. A *strict Lyapunov* function for the equilibrium x_0 of (7) is a real-valued, continuously differentiable function $V : U \subset R^n \rightarrow R$ defined on a neighborhood U of x_0 which satisfies:

a. *Minimum*. V has a minimum at x_0, i.e. $V(x) \geq 0$ *for all* $x \in U$ *and* $V(x) = 0$ *iff* $x = x_0$

b. *Decrease*. V is strictly decreasing along solution trajectories of (7) in U except for the equilibrium. A sufficient condition is $\dot{V} < 0$ *for all* $x \in U \setminus x_0$.

Thus, two important properties are deduced for a strict Lyapunov function [4]:

- If we have a strict Lyapunov function, then the equilibrium is asymptotically stable;
- Compact sublevel sets of a strict Lyapunov function are subsets of the basin of attraction of the equilibrium.

This chapter is organized as follows. The section 2 is dedicated to Lyapunov functions computational analysis. There are exposed the basic outlines of CLF concept and also the related outlines: LMI approach and SOS Lyapunov functions. In the section 3, a computational Lyapunov function is searched for the mixing flow dynamical system in a slightly perturbed form. After presenting the mathematical context of the 2d mixing flow dynamical system, together with recent results in the field, the results of searching a CLF for the mixing flow are presented. The section 4 is dedicated for conclusions and further aims in the topic. The chapter ends with references.

2. Computational Lyapunov stability analysis

2.1 Control Lyapunov functions

The concept of control Lyapunov function (CLF) is a very useful appliance in solving stability tasks. We search to stabilize a nonlinear system by selecting a

Lyapunov function $V(x)$ and then try to find a feedback control $u(x)$ that gives $\dot{V}(x, u(x))$ negative definite. If with an arbitrary choice of V this attempt may fail, when V(x) is a CLF, to find a stabilizing control law $u(x)$ is easier.

Similar with the system (5), we can define a control system like follows:

$$\dot{\boldsymbol{x}} = \boldsymbol{f}(\boldsymbol{x}, \boldsymbol{u}) \tag{13}$$

where $\boldsymbol{u} \in U \subset R^m$ is the control. We speak about an *open-loop control* if u is function of time, $\boldsymbol{u} = \boldsymbol{u}(t)$ and *closed-loop* if $\mathbf{u} = \mathbf{k}(\mathbf{x})$. The closed-loop control is in fact the *feedback control*. We speak also about *feedback stabilized system* if the feedback has been fixed, $\mathbf{u} = \mathrm{k}(\mathbf{x})$ and the equilibrium in the origin has a desired stability property.

The system (13) is called *locally, asymptotically null-controllable*, [4] if for every ρ in a neighborhood of the origin there is an open-loop control u such that the solution of the system with initial value ρ tends asymptotically towards the origin, i.e. if it is possible to steer the system state asymptotically to the origin. A *control Lyapunov function* (CLF) for such a system, introduced by Sontag 1983 [5], is a positive definite function V such that

$$inf_{\,u \in U} \nabla V(\boldsymbol{x}) \bullet \boldsymbol{f}(\boldsymbol{x}, \boldsymbol{u}) \leq -\gamma(\|\boldsymbol{x}\|) \tag{14}$$

where γ is a comparison function [4]. Asymptotic null-controllability cannot be characterized by smooth control Lyapunov functions and one must resort to more general definitions of differentiability like the Dini- or the proximal sub-differential [6].

For asymptotically null-controllable systems the equilibrium at the origin is sometimes referred to as weakly asymptotically stable, in contrast to strongly asymptotically stable equilibrium, where every choice of **u** leads to states being attracted asymptotically to the equilibrium.

As mentioned in the previous section, the stability concept has a lot of approaches: we have the classic Lagrange, Dirichlet and Lyapunov stability but, depending on the context, we also have input–output stability, hyperstability, input-to-state stability [3]. The last one, input to state stability (ISS) was introduced by Sontag [7] and it is interesting by the idea of characterizing a certain kind of stability at the origin imposing for the Lyapunov function V the condition:

$$\nabla V(\boldsymbol{x}) \bullet \boldsymbol{f}(\boldsymbol{x}, \boldsymbol{u}) \leq -\gamma(\|\boldsymbol{x}\|) + \alpha(\|\boldsymbol{u}\|) \tag{15}$$

where α and γ are comparison functions [4]. The origin is thus an asymptotically stable equilibrium of the system $\dot{\boldsymbol{x}} = \boldsymbol{f}(\boldsymbol{x}, 0)$ and a practically stable equilibrium of the system $\dot{\boldsymbol{x}} = \boldsymbol{f}(\boldsymbol{x}, \boldsymbol{u})$ for $\|\boldsymbol{u}\| \leq u_{max}$, with $u_{max} > 0$ a (not too large) constant. Moreover, the smaller u_{max} is, usually interpreted as a bound on the perturbation u, the closer solutions of the system will be to zero in the long run.

2.2 Stability analysis using sum of squares Lyapunov functions

The stability of dynamical systems is basically carried out by Lyapunov theory. For linear systems, the construction of an "energy-like function" – the Lyapunov function, fulfilling certain positivity conditions, is not difficult. For a system $\dot{x} = Ax$ this implies finding a matrix P such that $A^T P + PA$ is negative definite [8]. Then the associated Lyapunov function is given by $V(x) = x^T Px$. But although obviously, it was seen only recently that, in this context, both $V(x)$, $\dot{V}(x)$ are *sum of squares functions*!

Sum of squares (SOS) optimization is a quite new technique at the interface between convex optimization and computational algebra. Recently it had significant impact not only in optimization, but over several disciplines as well, especially in control theory. Besides the stability analysis of nonlinear systems, SOS has opened a new direction in approaching different types of systems and answering different analysis questions. The SOS technique generalizes a well-known computational appliance in linear robust control theory, "Linear Matrix Inequalities" – LMI. Parrilo and Ahmadi had important contributions in this field. [9] Using LMI in different analysis problems is advantageous, since there are efficient algorithms developed in the framework of semi-definite programming (SDP) [8, 9]. The SOS uses these types of algorithms, but all questions are formulated at polynomial level, or in polynomial-matrix terms.

Related to the Lyapunov functions, to construct them "by test" requires analytic skills of the researchers and moreover, depend on the small-state dimensions. When the vector field f and the Lyapunov function candidate V are both polynomial, the Lyapunov conditions are polynomial non-negativity conditions, which are quite hard to test. This could be one of the reasons for lack of efficiency in the algorithmic construction of a Lyapunnov function. But, if the non-negativity conditions are replaced by SOS conditions, then constructing the Lyapunov function can be done efficiently using semi-definite programming.

There are a lot of the control problems which follow the same two steps: i) recasting the primal problem as a Lyapunov-type problem and then, ii) constructing a sum-of-squares relaxation to the problem. We present in what follows a theoretical tool which is basic in the sum of squares computational approach [8].

Given $x \in R^n$ we denote the ring of multivariable polynomials with real coefficients by $R[x]$ and the subset of sum-of-squares polynomials in the variable x by $\sum[x]$. Sometimes it may be necessary to indicate the maximum degree of a polynomial or sum-of-squares polynomial in which case we use the subscript notation $R_d[x]$ or $\sum_d [x]$ where d is a positive integer.

Theorem 3. Let $x^* = 0 \in D \subset R^n$ an equilibrium point of (7). If there exists a function $V : D \to R$ continuously differentiable such that the following hold:

$$V(0) = 0; \tag{16}$$

$$V(\boldsymbol{x}) > 0, \forall x \in D \setminus \{0\} \tag{17}$$

$$\dot{V}(\boldsymbol{x}) \leq 0 \forall x \in D. \tag{18}$$

Then x^* is stable. Moreover, x^* is asymptotically stable if (11) holds.

Theorem 4. Let $x^* = 0$ be equilibrium for (7) and assume that $D \subset R^n$ is a given domain which includes x^*. Assume there exists a continuously differentiable function $V : D \to R$ and positive constants k_1, k_2, k_3 such that

$$k_1 \|x\|^p \leq V(x) \leq k_2 \|x\|^p \tag{19}$$

$$\dot{V}(x) \leq -k_3 \|x\|^p, p \in Z \tag{20}$$

for all $t \geq 0, x \in D$. Then x^* is exponentially stable. Furthermore, if the assumptions hold when $D = R^n$, then x^* is globally exponentially stable.

The following theorem illustrates how sum-of-squares programming can be used to construct a polynomial stability certificate, in this case a Lyapunov function, for an equilibrium point of (7). The domain D must be defined. In particular D must be representable as a semi-algebraic set. If we consider that the domain of interest D is represented by all points that satisfy $\beta(x) \leq 0, \beta \in R[x]$, then we have the following result [8].

Theorem 5. If there exists a polynomial function V, sum-of-squares polynomials r_1, r_2 and positive definite polynomial functions φ_1, φ_2 all of bounded degree, such that

$$V(x) + r_1(x)\beta(x) - \varphi_1(x)\ \epsilon\Sigma[x] \tag{21}$$

$$-\dot{V}(x) + r_2(x)\beta(x) - \varphi_2(x)\ \epsilon\Sigma[x] \tag{22}$$

then the equilibrium point x^* of (7) is asymptotically stable.

A few questions immediately come to mind: Do stable polynomial systems always admit a sum-of-squares Lyapunov function? How conservative are we being by limiting ourselves to positive polynomials that admit a sum-of-squares decomposition? Can we determine a priori the degree of the Lyapunov function required? The answer to the first question is simply, no.

There is a vast literature on the SOS method to compute Lyapunov functions in various settings and for different kinds of systems [4]. Between them, Parillo given important result on SOS and SDP (Semi Definite Problems) programming in finding Lyapunov functions [9]. As specified above, he introduced an efficient LMI program for finding Lyapunov functions as sum of squares for polynomial systems. The above setting is simplified for finding global Lyapunov functions, but if we are interested to find SOS Lyapunov functions on a compact domain, is important to mention the "Positivstellensazt" as useful appliance [9].

In the case of a dynamical system of a simple polynomial form, the stability study and Lyapunov function search can begin with a function test which facilitates the further analysis, as we see in the section 3.

3. Existence of control Lyapunov function for dynamical systems from excitable media

3.1 Recent results

The concerning for the dynamical models arising from excitable media is not new. The dynamical systems modeling the mixing flow have an important place, because of the complexity of the model. It is in fact about far from equilibrium class of models, with a very sensitive behavior to initial conditions.

Let us recall the statistical idea of a flow, generally represented by the application

$$\boldsymbol{x} = \Phi_t(\boldsymbol{X}), \boldsymbol{X} = \Phi_{t=0}(\boldsymbol{X}). \tag{23}$$

That means, X is mapped in x after a time t. In continuum mechanics, the relation (23) is named *flow*, it is a diffeomorphism of class C^k and must satisfy the relation

$$0 < J < \infty, J = \det\left(\frac{\partial x_i}{\partial X_j}\right), J = \det(D\Phi_t(x)) \tag{24}$$

where D denotes the derivation operation with respect to the reference configuration, in this case with **X**. If the Jacobean J is unitary, it is said we have an isochoric flow. The relation (24) implies two particles, $\mathbf{X}_1$ and $\mathbf{X}_2$ which occupy the same position **x** at a given moment, or a particle which splits in two parts. That means, non-topological motions like break up or disintegration *are not allowed.*

The mixing flow is a special type of flow, implying a basic fluid (water) in which a biological material is moving (mixing) in different conditions and with different velocities. Therefore, the *stretching and folding* are strongly related phenomena. With respect to X there is defined the basic measure of deformation, the *deformation gradient* $\boldsymbol{F}$ [10]:

$$\mathbf{F} = (\nabla_X \Phi_t(\mathbf{X}))^T, F_{ij} = \left(\frac{\partial x_i}{\partial X_j}\right). \tag{25}$$

For a material filament and correspondingly for a material surface, in a mixing flow, there are defined another two basic deformation measures, the *length deformation* λ and *surface deformation* η. In this context, a specific analysis for deformations of infinitesimal elements is the so-called "*good mixing concept*", related to the boundaries of the quantities λ and η. The class of flows with a special form of F is of very large interest in the literature, as it contains the so-called "constant stretch history motion" (CSHM flows). Details can be found in [10].

When studying the mixing flow phenomena, one starts from the widespread kinematic 2d mixing flow

$$\begin{cases} \dot{x_1} = Gx_2 \\ \dot{x_2} = KGx_1, \end{cases} \quad -1 < K < 1, G \in R. \tag{26}$$

Although this is a linear model, when associating the corresponding initial condition.

$$x_1(0) = x_1(t=0) = X_1; x_2(0) = x_2(t=0) = X_2 \tag{27}$$

it is obtained a complex solution for the Cauchy problem (26)-(27) [11]. From geometric standpoint, the streamlines of the above model satisfy the relation $x_2^2 - K \bullet x_1^2 = const.$ and this is corresponding to some ellipses with the axes rate $\left(\frac{1}{|K|}\right)^{1/2}$ if K is negative, and to some hyperbolas with the angle $\beta = \arctan\left(\frac{1}{|K|}\right)^{1/2}$ between the extension axis and x_2, if K is positive [10].

To this broad isochoric flow, we can associate easily the corresponding 3d dynamical system [11]:

$$\begin{cases} \dot{x}_1 = G \bullet x_2 \\ \dot{x}_2 = K \bullet G \bullet x_1, \quad -1 < K < 1, c = const. \\ \dot{x}_3 = c \end{cases} \tag{28}$$

with the third component for the moving velocity of the system.

In the 3d case, the non-periodic model exhibits a complicate behavior. A lot of comparative computational analysis proved the great influence of the parameters on the model behavior, leading to far from equilibrium models [11]. The perturbed model was also taken into account, and it was found out that its sensitivity with respect to the parameters is significant, both in 2d and 3d case [11, 12].

3.2 Existence of a CLF for the mixing flow dynamical system in a slightly perturbed form

The central aim in the study of the mixing flow dynamical system was associated rather with the fluid mechanics standpoint, namely analyzing the *efficiency of mixing* [10, 11].This is a concept which implies the analysis of deformation efficiencies in length and surface for the material mixed in the basic fluid. The physical phenomena associated are the *multiphase flow* phenomena, and the analysis and numeric simulation for these complex flows is in study. Briefly, in order to obtain a good behavior for the deformation efficiencies, the mathematical context must take into account the following three stages:

- modeling the global swirling streamlines;
- local modeling of the concentrated vorticity structure;
- introducing the elements of chaotic turbulence.

Because of its complexity, any perturbation on the mixing flow model changes the behavior of the model in a significant way. Therefore its stability is an important and challenging task. If in [13] it was found a SOS Lyapunov function working both for the initial and the feedback linearized form of the mixing flow dynamical system, this chapter brings the novelty of approaching the mixing flow dynamical system as a *polynomial differential system*, in order to get an optimal search for a Lyapunov function.

When taking into account the feedback control, the transformed model has a significant different repartition of the parameters in the model [12]. The stability analysis was started with the 2d case, with slight perturbations. In what follows, we consider the same slightly perturbed form of the 2d dynamical system as in [13], namely

$$\begin{cases} \dot{x_1} = Gx_2 + x_1 \\ \dot{x_2} = KGx_1 - x_2 \end{cases} - 1 < K < 1, G \in R. \tag{29}$$

For this model, it was found a strong result. A Lyapunov function V was found *both* for the initial model (29) and for its feedback linearized form. Namely, it was constructed the sum-of-squares function

$$V : R^2 \rightarrow R, V(\boldsymbol{x}) = x_1^2 + \frac{1}{|K|}x_2^2, \forall \boldsymbol{x} = (x_1, x_2) \in R^2. \tag{30}$$

The conditions (8)–(11) are fulfilled for suitable conditions on the parameters, for both models, so V is a Lyapunov function, and the origin is an asymptotically stable equilibrium point.

In what follows we consider the problem of finding a *control Lyapunov function* (CLF) for the model (29). We use a simplified form of the control system (5), namely we consider a control system of the form

$$\dot{\boldsymbol{x}} = f(\boldsymbol{x}) + g(\boldsymbol{x})u \tag{31}$$

where f and g are smooth vector fields, $x(t) \in R^n, u(t) \in R, f(0) = 0$.

Definition 2. A function V is a control Lyapunov function (CLF) for this system if $V : R^n \rightarrow R$ is a smooth, radially bounded, and positive definite function such that

$$inf_{u \in R}\left\{\frac{\partial V}{\partial \boldsymbol{x}}f(\boldsymbol{x}) + \frac{\partial V}{\partial \boldsymbol{x}}g(\boldsymbol{x})u\right\} < 0, \forall x \neq 0. \tag{32}$$

Existence of such a V implies that (29) is globally asymptotically stabilizable at origin.

We have to notice that the condition (32) for the CLF is in fact equivalent with that in (14) introduced by Sontag [6]. If such a CLF is given, it is shown [6] that a feedback low u = k(x) with k(0) = 0, can be constructed from the CLF producing a closed loop globally asymptotically stable. Hence, the problem of globally asymptotically stabilizing (31) is reduced to finding a CLF for the system.

Finding a CLF is a not trivial problem. It has been approached by multiple techniques, and Linear Programming, Positivstellensatz (P-sat) [14] and Linear Matrix Inequalities are only few examples. For the present aim the LMI approach [15, 16] helped to get better track of the model behavior.

A CLF must satisfy the inequality (32) which can be further evaluated as:

$$inf_{u\in R}\left\{\frac{\partial V}{\partial \boldsymbol{x}}f(\boldsymbol{x})+\frac{\partial V}{\partial \boldsymbol{x}}g(\boldsymbol{x})u\right\}=\begin{cases}-\infty & when\,\frac{\partial V}{\partial \boldsymbol{x}}g(\boldsymbol{x})\neq 0\\ \frac{\partial V}{\partial \boldsymbol{x}}f(\boldsymbol{x}) & when\;\frac{\partial V}{\partial \boldsymbol{x}}g(\boldsymbol{x})=0\end{cases}. \quad (33)$$

For a fixed x such that $\frac{\partial V}{\partial x}g(x)\neq 0$, we can make the inequality (32) hold by choosing a large value of u of the correct sign. Therefore, is essential to establish the set of x such that $\frac{\partial V}{\partial x}g(x)=0$. Thus, we want

$$\frac{\partial V}{\partial \boldsymbol{x}}f(\boldsymbol{x})<0\forall x\in R^n \; such\; that\; \frac{\partial V}{\partial \boldsymbol{x}}g(\boldsymbol{x})=0, x\neq 0. \quad (34)$$

Let us consider the 2d mixing flow dynamical model in its perturbed form (29). We have a polynomial system of differential equations, with monomials in the right hand sides of the system. In order to put it in a controlled form like (31), we take into account a simple control u at each of the equations of the model (29), namely

$$\begin{cases}\dot{x_1}=Gx_2+x_1-u\\ \dot{x_2}=KGx_1-x_2+u\end{cases} -1<K<1, G\in R. \quad (35)$$

The vector fields **f** and **g** are $f=\begin{pmatrix}Gx_2+x_1\\ KGx_1-x_2\end{pmatrix}, g=\begin{pmatrix}-1\\ 1.\end{pmatrix}$

We are looking for a positive definite symmetric matrix P, $P=\begin{pmatrix}P_{11} & P_{12}\\ P_{12} & P_{22}\end{pmatrix}$ and define a CLF function like

$$V=\frac{1}{2}\boldsymbol{x}^T P\boldsymbol{x}, \quad (36)$$

that means

$$V=\frac{1}{2}\left(x_1^2P_{11}+2x_1x_2P_{12}+x_2^2P_{22}\right)$$

such that (34) holds.

So, we search P by setting the condition $\frac{\partial V}{\partial \boldsymbol{x}}g(\boldsymbol{x})=0$ and studying when the condition $\frac{\partial V}{\partial \boldsymbol{x}}f(\boldsymbol{x})<0$ is fulfilled.

We have scalar inners in the condition (34). So, $\frac{\partial V}{\partial \boldsymbol{x}}g(\boldsymbol{x})=0$ implies

$$(P_{12}-P_{11})x_1+(P_{22}-P_{12})x_2=0.$$

From here we take x_2 function of x_1 and replace in $\frac{\partial V}{\partial \boldsymbol{x}}f(\boldsymbol{x})$. Thus, we obtain a quadratic form like follows:

$$\frac{\partial V}{\partial \boldsymbol{x}}f(\boldsymbol{x})=(P_{11}+KGP_{12})x^2+(P_0GP_{11}+P_0KGP_{22})x+(P_{12}-P_{22})P_0^2, \quad (37)$$

with the notation

$$P_0 = \frac{P_{12} - P_{11}}{P_{12} - P_{22}}.$$

The sign of the form (37) is of course dominated by the coefficient of x^2. Thus, we have

$$\frac{\partial V}{\partial \boldsymbol{x}} f(\boldsymbol{x}) < 0$$

when

$$P_{11} + KGP_{12} < 0. \quad (38)$$

Taking into account that K,G are also real parameters, the inequality (38) implies few tests. Therefore for the moment we choose the matrix P like

$$P = \begin{pmatrix} 1 & ½ \\ ½ & 1 \end{pmatrix}. \quad (39)$$

With this choice we find the condition

$$KG < -2, \quad (40)$$

which is a feasible condition since from the model we have $-1 < K < 1, G \in R$.

Now we can verify if the test value (39) of P can produce a feasible CLF function V. LMI approach has very reliable numeric tools [14, 15]. We choose for the present aim the basic matrix inequality, namely the Lyapunov inequality:

$$A^T P + PA \leq 0 \quad (41)$$

where A is the system matrix and P is the matrix in the definition (36) of CLF. In our model (35), the matrix A is given by:

$$A = \begin{pmatrix} 1 & KG \\ G & -1 \end{pmatrix}, \quad (42)$$

and thus we obtain

$$A^T P + PA = \begin{pmatrix} 2P_{11} + 2P_{12}KG & P_{11}G + KGP_{22} \\ P_{11} + P_{22}KG & 2P_{12}G - 2P_{22} \end{pmatrix}. \quad (43)$$

Replacing with the test values from (39) for P, the condition "$A^T P + PA$ negative semidefinite" implies the following conditions for the parameters

$$KG < -2, \quad (44)$$

$$G < -\frac{1}{2} \bullet \frac{1}{1+K}.$$

These conditions are feasible, for a negative G.

Thus we can conclude that, in feasible conditions for the parameters, the model (35) can admit a CLF V and thus the model can be globally stabilizable at the origin.

4. Conclusions

In the present chapter, the aim of stabilizing the two-dimensional dynamical systems arising from excitable media is taken into account.

Stabilizing a dynamical system from excitable media is not an easy task, especially because of the sensitive dependence of these models on initial conditions. The mixing flow models are not an exception, since the repartition of their parameters has a great influence on the model trajectory behavior.

The Lyapunov function is a strong appliance in studying stability. Still, constructing it is not easy. This is why approaching this for controllable systems make easier the task of searching the stability.

Searching for a Lyapunov function becomes easier if we take into account the iterative algorithm in the "sum of squares" programming. For the mixing flow example of this chapter, the test values (39) for the matrix P shown that in feasible conditions for the parameters, we can construct a CLF for the model (35) and thus the model is globally stabilizable in the origin.

Constructing the CLF functions for controllable systems provide interesting results, and this is shown by the above 2d mixing flow model. The 2d mixing flow model is controllable in some slightly perturbed forms [12, 13]. Therefore, in this chapter the standpoint is that we started with the simple control u taken on both equations of the dynamical system. The control is not implied in the further calculus on finding the CLF V, which made easier the above analysis. A next aim is to find CLF functions for other perturbations cases of the mixing flow model, on one hand, and also to approach iterative LMI algorithms for finding the matrix P. The test values (39) for the matrix P give rise to an easier further statement of a SDP problem for the mixing flow dynamical system.

Thus, this chapter adds a new standpoint in the mathematical context of the mixing flow dynamical system, by considering it as a polynomial system of differential equations. For a model whose mathematical apparatus is rather associated with mechanics and fluid mechanics issues, approaching the stability by searching Lyapunov functions with computational algebraic appliances is new. This approach could open a new way in the study of stabilizing of the mixing flow dynamical model - which is a far from equilibrium model, and also in the qualitative analysis of the differential systems associated to transport phenomena.

Author details

Adela Ionescu
University of Craiova, Romania

*Address all correspondence to: adelajaneta2015@gmail.com

References

[1] Henson, M. A. and Seborg, D. E., eds, Nonlinear Process Control. Prentice Hall PTR, 1996

[2] Krasovskii, N, Stability of Motion. Applications of Lyapunov's second method to differential systems and equations with delay. Trans. by J. L. Brenner. Stanford University Press, 1963. Originally published in Russian by Fizmatgiz, Moscow, 1959

[3] Isidori, A, Nonlinear Control Systems. Springer Verlag, 1989. DOI: 10.1007/978-3-662-02581-9

[4] Giesl, P, Hafstein,S, Review on computational methods for Lyapunov functions. Discrete and Continuous Dynamical Systems SeriesB, 2015, 20 (8); 1–41. DOI: 10.3934/dcdsb.2015.20

[5] Sontag, E, A Lyapunov-like characterization of asymptotic controllability. SIAM J. of Control and Optimization, 1983, 21(3), 462–471

[6] Sontag, E, Mathematical Control Theory. Deterministic finite dimensional systems. 2nd editon, Springer 1998. ISBN: 978–0387984896

[7] Sontag, E, A Lyapunov-like characterization of asymptotic controllability. SIAM J. of Control and Optimization, 1983, 21(3), 462–471

[8] Anderson, J, Papadimitriou, A, Advances in computational Lyapunov analysis using sum-of-squares programming. Discrete and Continuous Dynamical systems series B, vol 20(8), 2015; 2361–2381. DOI:10.3934/dcdsb.2015.20.2361

[9] Ahmadi, A. A. and Parrilo, P. A, Sum of squares certificates for stability of planar, homogeneous and switched systems. IEEE Transactions on Automatic Control, 2017 62(10), 5269–5274. DOI: 10.1109/tac.2016.2647253.

[10] Ottino, J, The Kinematic of Mixing: Stretching, Chaos and Transport. Cambridge Texts in Applied Mathematics. Cambridge University Press, 1989

[11] Ionescu, A, Recent Trends in Computational Modeling of Excitable Media Dynamics. Lambert Academic Publishing, 2010

[12] Ionescu, A, New qualitative features for the 2d dynamical system associated to a mixing flow model. Atti della Accademia Peloritana dei Pericolanti, 2017, 95(2), A2–1-A2–7. DOI:10.1478/AAPP.952A2

[13] Ionescu, A, Stability conditions for the mixing flow dynamical system in a perturbed version. Atti della Accademia Peloritana dei Pericolanti, 2019, 97(2); A5–1-A5–10. DOI: 10.1478/AAPP.972A5

[14] Papadimitriou, C.H, Steiglitz, K, Combinatorial Optimization: Algorithms and Complexity, Prentice Hall, Englewood Cliffs, New Jersey, 1982

[15] Boyd, S, Ghaoui L.El, Feron, E, Balakrishnan, V, Linear Matrix Inequalities in System Control Theory, SIAM Studies in Applied Mathematics, Philadelphia PA, vol 15, 1994. ISBN: 0–89871-334-X

[16] VanAntwerp, J. G, Braatz, R.D, A tutorial on linear and bilinear matrix inequalities. J. of Process Control, 2000 (10), 363–385